A GUIDE TO THE NIGHT SKY

Dr Sara Webb

With foreword by
Dr Kirsten Banks

A GUIDE TO THE NIGHT SKY

The science and lore behind our constellations

ILLUSTRATIONS BY

Aidan Meighan

Smith Street Books

CONTENTS

Foreword ... 6
Introduction ... 10
How to view the night sky ... 12
... And what you might see ... 14
Glossary ... 198

ANCIENT CONSTELLATIONS
18

Andromeda ... 20
Aquarius ... 22
Aquila ... 24
Ara ... 26
Aries ... 28
Auriga... 30
Boötes ... 32
Cancer ... 34
Canis Major ... 36
Canis Minor ... 38
Capricornus ... 40
Cassiopeia ... 42
Centaurus ... 44
Cepheus ... 46
Cetus ... 48
Corona Australis ... 50
Corona Borealis ... 52
Corvus ... 54
Crater ... 56
Cygnus ... 58
Delphinus ... 60
Draco ... 62
Equuleus ... 64
Eridanus ... 66
Gemini ... 68
Hercules ... 70
Hydra ... 72
Leo ... 74
Lepusv ... 76
Libra ... 78
Lupus ... 80
Lyra ... 82
Ophiuchus ... 84
Orion ... 86
Pegasus ... 88
Perseus ... 90
Pisces ... 92
Piscis Austrinus ... 94
Sagitta ... 96
Sagittarius ... 98
Scorpius ... 100
Serpens ... 102
Taurus ... 104
Triangulum ... 106
Ursa Major ... 108
Ursa Minor ... 110
Virgo ... 112

MODERN CONSTELLATIONS

114

Antlia ... 116
Apus ... 118
Caelum ... 120
Camelopardalis ... 122
Canes Venatici ... 124
Carina ... 126
Chamaeleon ... 128
Circinus ... 130
Columba ... 132
Coma Berenices ... 134
Crux ... 136
Dorado ... 138
Fornax ... 140
Grus ... 142
Horologium ... 144
Hydrus ... 146
Indus ... 148
Lacerta ... 150
Leo Minor ... 152
Lynx ... 154
Mensa ... 156
Microscopium ... 158
Monoceros ... 160
Musca ... 162
Norma ... 164
Octans ... 166
Pavo ... 168
Phoenix ... 170
Pictor ... 172
Puppis ... 174
Pyxis ... 176
Reticulum ... 178
Sculptor ... 180
Scutum ... 182
Sextans ... 184
Telescopium ... 186
Triangulum Australe ... 188
Tucana ... 190
Vela ... 192
Volans ... 194
Vulpecula ... 196

FOREWORD

As a proud Wiradjuri woman and astrophysicist, I see the night sky not only through the lens of science but also through the profound cultural teachings of my people.

For tens of thousands of years, humans have gazed up at the night sky, finding meaning and wonder in the seemingly chaotic sprawl of stars. The practice of identifying patterns in the stars – which we call constellations – and connecting them to stories, lessons and knowledge is a deeply human trait. Cultures across the world have used these patterns as navigational aids, to mark the changing seasons and to reflect on the values and lore of their communities. This book is part of that grand human tradition, inviting readers to explore the constellations and their stories. It is also an opportunity to widen our perspective and appreciate the astonishing diversity of ways people have connected with the stars, ways that often go beyond the constellations of modern astronomy.

As a proud Wiradjuri woman and astrophysicist, I see the night sky not only through the lens of science but also through the profound cultural teachings of my people. Indigenous Australians, including the Wiradjuri, have one of the oldest continuous astronomical traditions in the world. It stretches back tens of thousands of years. Our ancestors studied the sky with deep curiosity and care, mapping the movements of stars and planets, observing the Moon's phases and paying close attention to phenomena such as eclipses, comets and even aurorae. These observations were not only scientific, but interwoven with culture, spirituality and practical knowledge. The stars, after all, have long served as guides, not just for navigating physical landscapes but also for navigating human connections and community.

In Aboriginal Astronomy, constellations take many forms. Some are collections of stars joined together in patterns that tell stories or depict significant cultural figures, animals or events, much like the constellations catalogued in this book. Others are single stars that represent specific people, objects or creatures of cultural importance. Most strikingly, there are the 'dark constellations': patterns that emerge not from the stars themselves but from the dark spaces between them.

One of the most famous examples, and my favourite constellation, is the celestial emu. Known as Gugurmin in the Wiradjuri language, this figure stretches across the Milky Way. Its head is the dark patch known as the Coalsack Nebula, near the Southern Cross, while its body is formed from the centre of the Milky Way galaxy, and its legs trail beyond. The celestial emu is not just a story; it is a seasonal marker, guiding communities on when to hunt for emu eggs. Such practical applications demonstrate the deep integration of astronomy into daily life.

Globally, Indigenous cultures have developed their own unique ways of interpreting the night sky. Polynesian peoples used the stars to navigate vast ocean distances, and stars have been used as markers of the changing seasons across multiple global cultures. In Inuit astronomy, the stars are woven into stories of survival, resilience and guidance. Some represent hunters and animals, and others depict individuals who overcame hardship and became celestial beings. These narratives not only explain the cosmos, but pass down essential knowledge of the land, seasons and wayfinding, highlighting the deep connection between Indigenous star lore and both practical and spiritual wisdom.

These perspectives remind us that astronomy is not solely about understanding the physical universe: it is also about understanding our place within it. By weaving survival strategies and cultural practices into their celestial interpretations, Indigenous peoples worldwide demonstrate a holistic approach to knowledge that resonates deeply with the human experience.

One of the most remarkable aspects of Indigenous Astronomy is its practicality. Constellations were not just stories for entertainment, they were vital tools for survival. They served as calendars, indicating the best times to plant, harvest and hunt. The night sky was also a navigation system, helping travellers find their way across vast landscapes. These celestial maps were precise, passed down through generations with incredible accuracy, and deeply ingrained in oral traditions.

In Aboriginal cultures, knowledge was often encoded in songlines – oral traditions that linked specific places, stories and star patterns. These songlines were not only mnemonic devices but also moral and legal codes, teaching the lore and law of a community. Through stories about the stars, children learned about the consequences of actions, the importance of cooperation, and their responsibilities to the land and to each other. These lessons remain relevant today, as we grapple with the challenges of living sustainably on this planet.

Modern astronomy, with its set of eighty-eight constellations rooted in Greek and Roman traditions, provides a valuable framework for mapping the sky. Names like Orion, Leo and Cassiopeia are familiar to stargazers worldwide. Yet these constellations represent only one perspective among many. Indigenous star knowledge offers alternative ways of seeing and interpreting the cosmos, enriching our understanding and appreciation of the night sky. By exploring these perspectives, we are reminded that science and culture are not separate domains but intertwined ways of making sense of the world.

The growing recognition of Indigenous Astronomy, both in Australia and globally, highlights its enduring value. Indigenous knowledge has contributed to modern science in remarkable ways, from aiding in the understanding of variable stars to providing oral records of supernovae and eclipses. Embracing these perspectives not only honours the wisdom of those who came before us but also enhances our collective ability to explore and understand the universe.

This book is a celebration of constellations, inviting you to trace the patterns of the night sky and uncover the stories behind them. As you read, I encourage you to think not only about the constellations themselves but about the broader human tradition of stargazing. Consider the countless generations of people who have looked up at these same stars, finding guidance, inspiration and meaning. Reflect on the ingenuity and creativity required to map the heavens and connect them to life on Earth. And remember that the night sky is a shared heritage. It belongs to all of us.

As we look to the future, it is vital to continue this tradition of looking up. The stars remind us of our place in the universe, of the vastness of space and time, and of the interconnectedness of all things. They challenge us to think big, to dream and to explore. Whether we are tracing the constellations of modern astronomy, uncovering the ancient stories of Indigenous cultures, or marvelling at the distant galaxies revealed by powerful telescopes, the act of stargazing connects us to something larger than ourselves.

I hope this book inspires you to step outside on a clear night and look up. Take the time to find some of the constellations described in these pages. Imagine the people who named them and the stories they told. Seek out the dark constellations, those patterns that emerge from the spaces between the stars. And remember that the night sky is not just a map of the universe but a map of human imagination and connection. May it guide you, as it has guided so many before us, to new discoveries and deeper understanding.

Dr Kirsten Banks

INTRODUCTION

Welcome to the night sky.
It's been waiting for you!

We have always been fascinated by the stars. The night sky offers us a glimpse beyond our Earth-bound lives and connects us directly to the cosmos. As an astronomer, I get to think about the stars, and beyond, almost every day.

My own fascination with the sky began when I was young, staring out the car window as we drove out of the city to visit Nan. I watched in awe as the stars came alive the further away from the city we got. I still look up the night sky and marvel at everything I know is out there in the Universe.

On a very clear night, away from city lights, you might be lucky to spot 2000 stars, twinkling away. Out of the sea of stars, there are typically fifty-eight that are very important for navigators. Across the years, many different nautical almanacs have listed the brightest stars in the night sky, right across the globe, and offered information for their use in navigation. Before the introduction of modern technology, sailors and aviators were able to use these almanacs, along with instruments known as sextants, to work out exactly where they were. As you can imagine, this was extremely helpful. It is a skill that some maritime and aviation professionals still need to know, just in case the technology fails.

Astronomers get to know the stars fairly well, especially the ones that help guide our way through the cosmos and that mark the interesting areas in our Universe that we are exploring.

However, you don't have to be an astronomer to know your way around the stars. *A Guide to the Night Sky* is your companion to the eighty-eight officially recognised constellations. It explains how they came to be and tells you what you might see in them, with or without binoculars or telescopes. This book is your guide to the sky, giving you background into the development of constellations throughout time, the objects within them and the constellations themselves. You will learn the history and lore of each constellation, how to find it in the night sky, the brightest star within and the deep-sky objects it contains.

I hope
A Guide to the Night Sky
ignites
your passion
for exploration
and discovery.

HOW TO VIEW THE NIGHT SKY ...

Before you head out to your backyard to search for a constellation, there are a few concepts and tips that will help you find what you are looking for.

Where to look

✦ **CELESTIAL SPHERE**

The celestial sphere is an imaginary sphere around the Earth that is used to describe the positions of celestial objects, including the constellations.

✦ **CELESTIAL EQUATOR**

The celestial equator is an imaginary line in the sky that extends from Earth's equator.

✦ **NORTHERN AND SOUTHERN SKIES**

The region of the sky to the south of the celestial equator is called the Southern Sky, and to the north is the Northern Sky.

✦ **NORTHERN, SOUTHERN AND EQUATORIAL CONSTELLATIONS**

In this book, each constellation is described as northern, southern or equatorial. Northern constellations sit in the Northern Sky, and are easier to see if you are in the Northern Hemisphere, and vice versa for Southern constellations. Equatorial constellations sit across the celestial equator, which makes them visible from most places on Earth.

Twelve of the constellations are also known as the Zodiacs. Each of the Zodiacs is associated with a particular time of the year.

Along with the best viewing month for each constellation, this book explains the places on Earth where it can be seen from. These positions are described using latitudes.

✦ **LATITUDE (AND LONGITUDE)**

Latitude and longitude are imaginary lines that help us accurately describe locations on Earth's surface.

Lines of latitude run parallel to the equator. These are used in this book to describe where on Earth a particular constellation can be seen from. They are measured in degrees, and are positive if they are north of the Equator and negative if they are south of it. The Equator itself is at 0 degrees. The North Pole is at 90 degrees N (+90°) and the South Pole is at 90 degrees S (–90°).

(We don't need to use lines of longitude in this book, but they run from the North Pole to the South Pole, and are sometimes called meridians.)

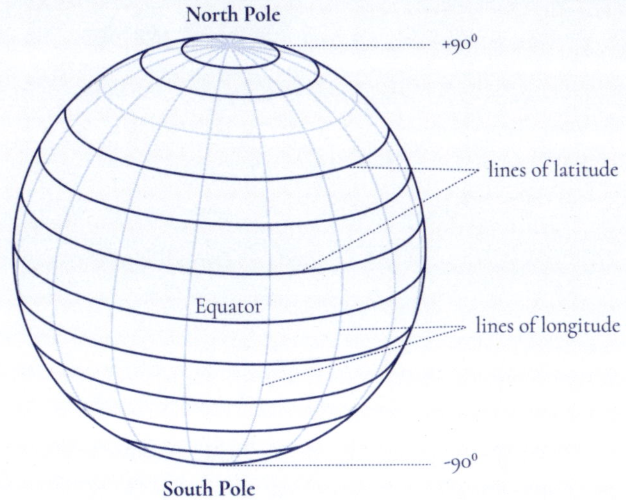

How to look

✦ **NAKED EYE**

You can start stargazing with just your eyes and a dark sky. To get the best view, choose a moonless night and escape the bright lights of the city.

✦ **BINOCULARS**

If you want to take the next step, use a pair of small to medium binoculars to explore the night sky. Binoculars magnify your vision and allow you to zoom in on nearby galaxies and planetary nebulae.

✦ **TELESCOPES**

If you are (or become!) a keen sky explorer, a telescope is the best option. But, before investing in a large or expensive setup, I encourage you to join or participate in an amateur astronomy club. There are several different types of telescopes, each suited to different needs. You can learn using other people's equipment, which will also help you decide which type is best for you.

✦ **NIGHT-SKY APPLICATIONS**

No matter how you are stargazing, a night-sky application will quickly become your best friend. There are many available apps that will help you search the stars. Some, such as Stellarium, can be used on a computer or phone. Using an app that draws a constellation's artwork on the sky will help you learn the patterns of the stars. Eventually, you will be able to identify them without the help of the app. A night-sky app will also guide you to the exact position of a specific object within the constellation you are exploring.

... AND WHAT YOU MIGHT SEE

Once you can find the constellations, it's time to look within and around them. The following information explains what some of the amazing objects in the night sky are.

Types of stars

There are more stars in the universe than you can fathom. The most recent estimates suggest there are 400 billion stars in the Milky Way alone. To put that into perspective, with the naked eye (on a very dark and clear night, far away from any light pollution) we can see around 4500 stars, across the whole sky. That is less than 0.00000001% of our own galaxy.

Astronomers have classified the stars in our universe into seven groups based on their colour, which indicates their temperature. Blue O-type stars are the hottest, and red M-type stars are the coolest (relatively speaking – they are still very hot!).

Stars are also classified by size. In general, the larger a star is, the brighter it appears in the night sky. Other stars are very bright because of the materials that they are made of.

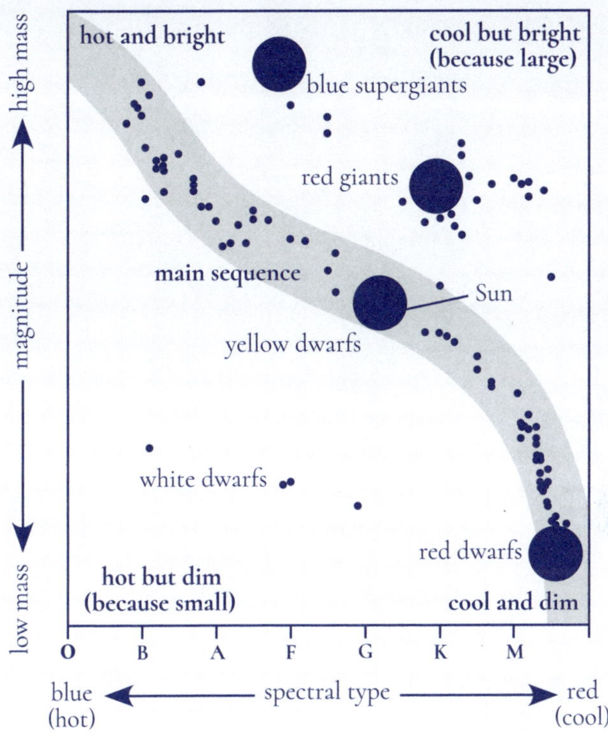

CLASSIFICATION OF STARS

✦ In the middle of the diagram, indicated by the silver area, are the main sequence stars. Ninety per cent of stars in the universe are classified as main sequence stars, but they won't stay that way forever. Eventually, the low-to medium-mass main sequence stars (like the Sun) will become red giants. Later still, they will be compressed into much smaller white dwarfs. High-mass main sequence stars have a different life cycle; they become red supergiants, then explode as supernovae and form either a neutron star or a black hole.

A fact that never ceases to amaze me is that seventy per cent of the stars in our galaxy are M-type red dwarfs. They are so faint that we can't see a single one of them with the naked eye.

Another important thing to know about stars is that a lot of them aren't alone. At least one-third of all stars are part of a binary (or triple or quadruple) star system, where two (or three or four) stars are gravitationally bound to each other. They often appear as a single star, until you look at them through a telescope.

Constellations

Constellations are the product of perspective and consensus. Throughout human history, many civilisations and cultures around the world had, and still have, their own interpretation of the night sky. They grouped visible stars into patterns and stories, which they used as a guide to the natural world that their survival depended upon.

The first catalogue of the stars was drawn up by the Greek astronomer Hipparchus in 129 BCE. Over the next 2000 years, many more star catalogues and celestial globes were made, identifying different groups of stars and assigning different names and stories to them.

By the 19th century, Europe had entered a new era of astronomy and exploration of the cosmos, and in 1922, the International Astronomical Union (IAU) held its first meeting in Rome, Italy. At that meeting, it was decided that there should be an agreed list of constellations. Eighty-eight were officially recognised, along with their Latin names. These constellations are now used by astronomers and stargazers across the world.

Each of the eighty-eight constellations is defined by its main stars and the specific area on the celestial globe that marks its boundaries. Together, they cover the entire celestial sphere. But, remember, the stars are constantly moving – the constellations we see today won't always look like this. They are only a snapshot in our galaxy's existence.

Asterisms

An asterism is a pattern or shape in the stars that is not directly related to the constellations but is still widely recognised. Two examples are the Big Dipper and the Summer Triangle. You can make up your own asterisms using the stars you see in the night sky.

Deep-sky objects

This book includes lists of deep-sky objects – within our galaxy, the Milky Way, and beyond – that can be seen in each constellation. They are often referred to by their official names, which begin with initials such as M (for Messier objects, which were first catalogued by French astronomer Charles Messier in 1781) and NGC (objects listed in the 1888 *A New General Catalogue of Nebulae and Clusters of Stars*).

✦ GALAXIES

There are over 2 trillion galaxies in our universe. We can see some of the nearby ones with the help of telescopes. Galaxies come in different shapes and sizes, including spiral galaxies like ours. You can imagine the Milky Way as a pinwheel, with spiral arms that spin around the galactic centre (the point that all the objects in our galaxy orbit). Right at the middle of our galactic centre is a supermassive black hole, Sagittarius A*, which has a mass 4.3 million times greater than the Sun. Other galaxies include elliptical galaxies, which appear to be round with no distant arms or features; and irregular galaxies, which are often the result of two or more galaxies merging.

✦ LARGE MAGELLANIC CLOUD

The Large Magellanic Cloud (LMC) is a dwarf galaxy that sits near the Milky Way – and by near, I mean 163,000 light-years away. It is one of our closest intergalactic neighbours. The LMC is roughly 1/100 the size of our galaxy. It is visible to the naked eye on clear nights in parts of the Southern Sky. The LMC is expected to merge with the Milky Way in 2.4 billion years.

✦ SMALL MAGELLANIC CLOUD

The Small Magellanic Cloud (SMC) is another dwarf galaxy, located 200,000 light-years from the Milky Way. The SMC has a mass approximately 7 billion times greater than the Sun, and is also visible to the naked eye on clear nights in parts of the Southern Sky.

✦ PLANETARY NEBULAE

When stars similar to the Sun come to the end of their lives, they expand and push off their outer layers. The gas and dust form ring or bubble-like structures that are shaped like planets.

✦ SUPERNOVA REMNANTS

Supernova remnants are nebulae created by the explosion of large stars.

✦ GIANT MOLECULAR CLOUD

A giant molecular cloud is an incredibly large and cold area of space that is mostly made up of molecular hydrogen. This is where most stars are first formed.

✦ OPEN CLUSTERS

Open clusters are made up of tens to thousands of stars that were formed from the same giant molecular cloud and remain loosely bound together. They are often found in spiral and irregular galaxies – as we live in a spiral galaxy, there are plenty to see in our night sky.

✦ GLOBULAR CLUSTERS

Globular clusters are large spherical groupings of stars that are tightly bound together by gravity. They can consist of millions of stars, and are associated with all types of galaxies. Globular clusters are often home to the oldest stars in a galaxy, and there are many to be seen in the Milky Way.

We have always looked to the sky to guide and inspire us. For millennia, stories have been spun from the stars that have taken root in our hearts. Many ancient constellations transcend cultures and time. They often revolve around the brightest stars in the sky – the ones that are easily seen with the naked eye.

Today, forty-eight constellations are classified as 'ancient'. Within these are the famous Zodiacs, which have their origins in ancient Babylonian astronomy and are associated with the twelve months of the year. Although this astronomer isn't really into astrology, she will disclose that she's a Taurus. Make of that what you will.

Over the centuries, some of the oldest constellations have been broken up into smaller ones and are now classified as 'modern'. This famously includes the grand ship of Greek mythology, which was sailed by the Argonauts on their search for the Golden Fleece. Argo Navis is now divided into Carina, Puppis and Vela.

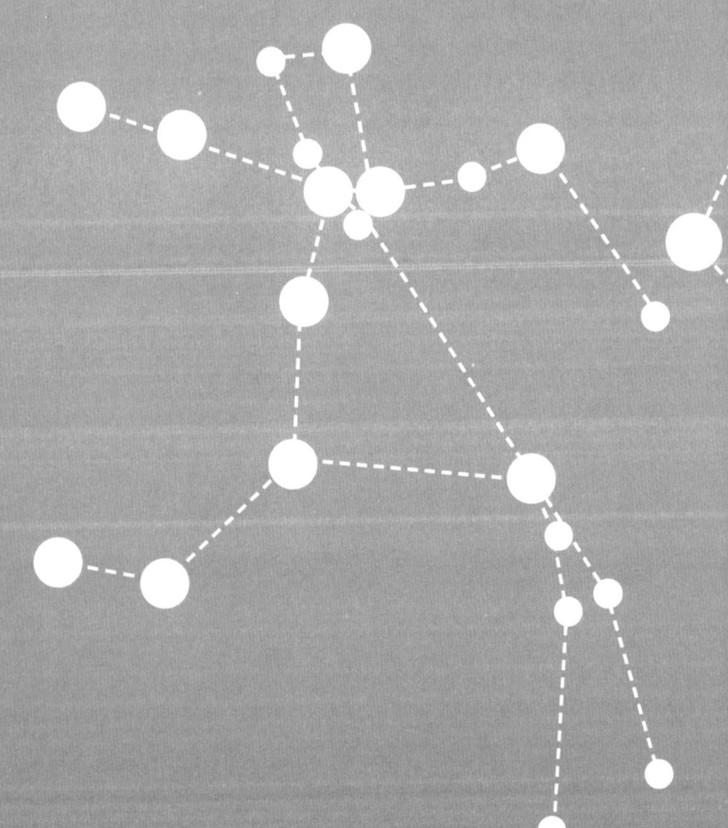

ANCIENT CONSTELLATIONS

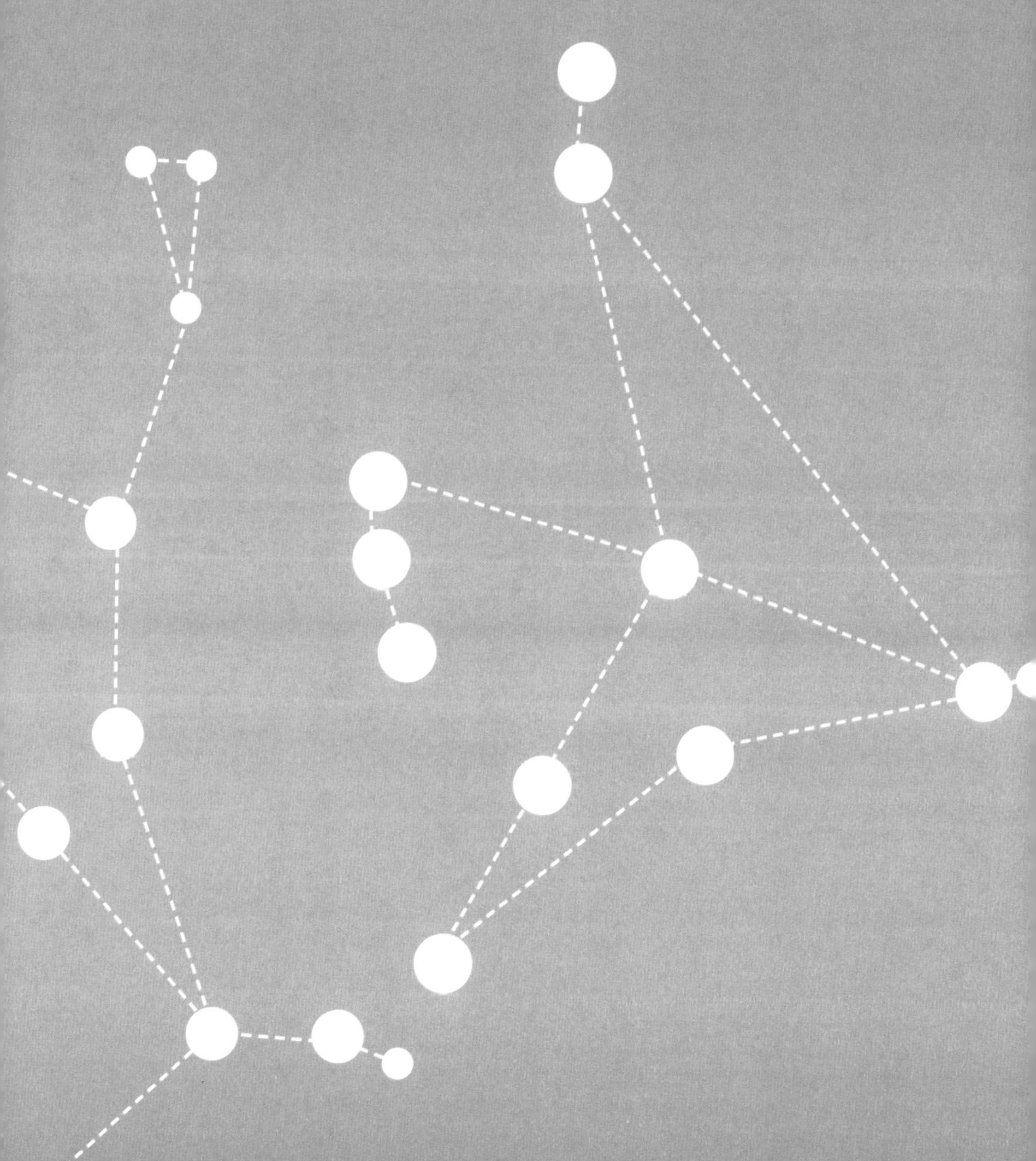

ANDROMEDA

The Chained Woman

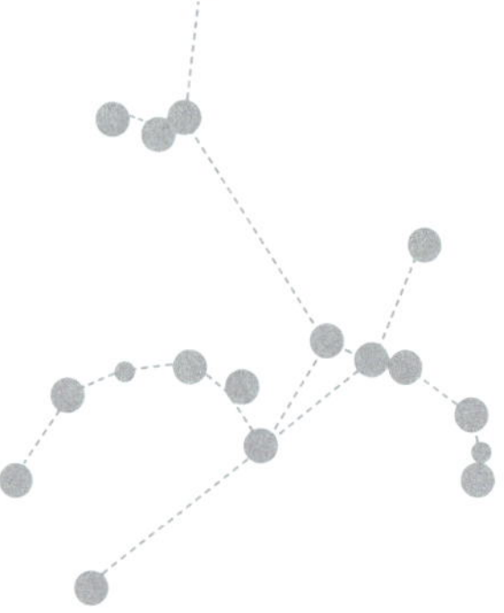

Lore

In Greek mythology, Andromeda was the daughter of King Cepheus and Queen Cassiopeia. Her mother bragged that she and her daughter were both stunningly beautiful, more beautiful than the famous sea nymphs. When Poseidon heard this, he sent the sea monster Cetus to destroy Cepheus's kingdom. His subjects believed they could save their country and the people by sacrificing Andromeda. Her parents chained her to the rocks by the sea and waited. However, Perseus flew past on Pegasus and fell in love with Andromeda. The hero killed Cetus and took Andromeda away to marry her.

How to see it

Andromeda is a northern constellation, visible at latitudes between +90° and -40° and best viewed in November. Find the square shape that makes up the body of Pegasus (p. 88). Andromeda's brightest star, Alpheratz, marks one corner of Pegasus, and the rest of the constellation extends from there like a tuning fork.

Deep-sky objects

The Andromeda Galaxy (M31) is the closest large galaxy to Earth. It is slowly moving towards the Milky Way – the two galaxies are expected to collide in 5 billion years.

Brightest star

Alpheratz is a binary star system whose two stars are in a close orbit. The brighter star is a rare mercury–manganese star.

Area and main stars

722 square degrees with 16 main stars

AQUARIUS

The Water Bearer

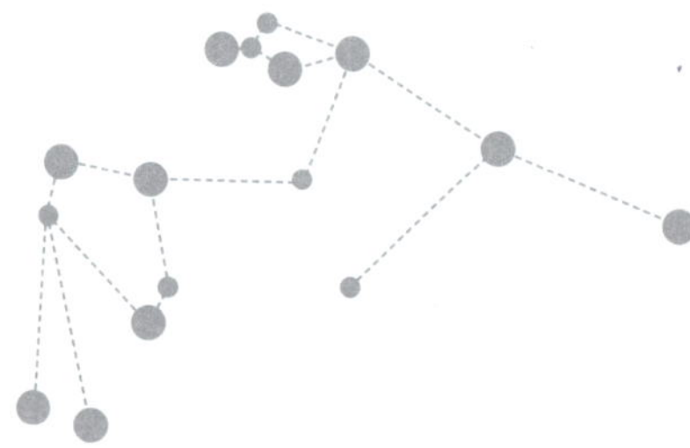

Lore

Aquarius is in a section of the sky where most constellations have water-related mythologies. In one story, Aquarius represents a young shepherd, Ganymede, who was abducted by the Greek god Zeus and taken to Mount Olympus to serve as a cupbearer to the gods. In Ancient Egypt, Aquarius was associated with the annual flood of the Nile. The myth was that the river would begin to flood when he dipped his jar into it.

How to see it

Aquarius is an equatorial constellation, visible at latitudes between +65° and -90° and best viewed in October. Although it is the tenth-largest constellation, it can be difficult to spot because its stars are quite faint. It is best seen on a moonless night, ideally away from light pollution. Using a night sky-application, familiarise yourself with the brightest star in Aquarius, Sadalsuud, which makes up the left shoulder of the water bearer.

Deep-sky objects

Aquarius is home to many brilliant deep-sky objects, including the Helix Nebula, one of the brightest and closest stellar nebulae to Earth.

Brightest star

Beta Aquarii, or Sadalsuud, is a yellow supergiant. Sadalsuud is almost 5 times the mass of the Sun but has expanded to 48 times its size. In some older mythologies, Sadalsuud marks the rising of the Sun at the end of winter. Many myths of luck and fortune are connected to this star.

Area and main stars

980 square degrees with 10 main stars

AQUILA

The Eagle

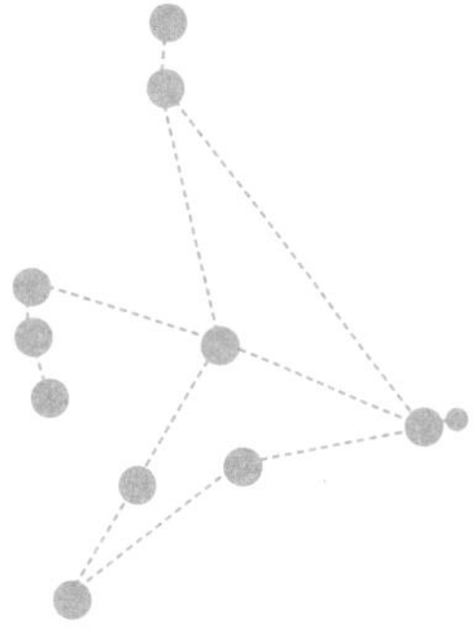

Lore

Aquila is the mighty eagle, once the king of birds. In Greek mythology, Aquila ran errands for Zeus and was a symbol of his power. One myth tells of Aquila delivering Zeus's thunderbolts during the war against the Titans. In another myth, Aquila was ordered to kidnap a Trojan prince, Ganymede, and take him to the top of Mount Olympus.

How to see it

Aquila is an equatorial constellation, visible at latitudes between +90° and -75° and best viewed in August. It is above Sagittarius (p. 98). In dark skies, it can be seen in front of the dust clouds of the Milky Way.

Deep-sky objects

Multiple spectacular deep sky objects can be seen here with binoculars or telescopes. One is NGC 6760, an open cluster of stars that formed around 100 million years ago. The M56 globular cluster is also visible to the aided eye, as is the colourful planetary nebula NGC 6781.

Brightest star

Altair is an A-type star that is the twelfth-brightest in the night sky. It is 1.86 times the mass of the Sun but is only 100 millions years old. When Altair was formed, the dinosaurs still ruled Earth.

Area and main stars

652 Sq degrees with 10 main stars

ARA

The Altar

Lore

Ara is one of the forty-eight constellations that were well known to observers for centuries and first catalogued by the Greek astronomer Ptolemy (100–170) in the 2nd century. It was originally referred to as an altea – an incense vessel or an altar with burning incense. It was associated with a myth written by the 1st-century Roman poet Marcus Manilius, in which Ara is the altar where Zeus and his fellow gods met and made vows before going into battle and defeating the Titans. It was named Ara in the 18th century.

How to see it

Ara is a southern constellation, visible at latitudes between $+25^{\circ}$ and -90° and best viewed in July. Locate Scorpius (p. 100). Trace the stars that form Scorpius and focus on the scorpion's tail, which appears to curve to one side. Ara sits beneath the curve of the scorpion's tail.

Deep-sky objects

This constellation has some notable objects that are best seen with larger telescopes, including the Stingray Nebula and the globular clusters NGC 6362 and NGC 6397.

Brightest star

Beta Arae is a young K-type supergiant. It has expanded to 141 times the size of the Sun and is 5800 times more luminous. It is estimated to be 50 million years old.

Area and main stars

237 square degrees with 8 main stars

ARIES

The Ram

Lore

Aries is associated with the Greek myth of the Golden Fleece. This epic story starts with Princess Ino's jealousy of her stepson Phrixus, the heir to the throne of Orchomenus. Ino convinced her husband, King Athamas, that he had been cursed and needed to sacrifice Phrixus. Zeus saw this and ordered Hermes to send a golden-fleeced ram to rescue Phrixus. Zeus later placed the ram in the sky.

How to see it

Aries is an equatorial constellation, visible at latitudes between +90° and -60° and best viewed in December. Stargazers in the Southern Hemisphere can spot it just above the northern horizon from spring to midsummer. The brightest star of Aries is at the point that forms part of the ram's head.

Deep-sky objects

Within Aries, there are a few easy-to-spot galaxies, including the spiral galaxies NGC 772 and NGC 972.

Brightest star

Alpha Arietis, or Hamal, is the fiftieth-brightest star in the night sky. It is a giant star that is 3.4 billion years old and has one known planet, which is 1.8 times the mass of Jupiter.

Area and main stars

441 square degrees with 4 main stars

AURIGA

The Charioteer

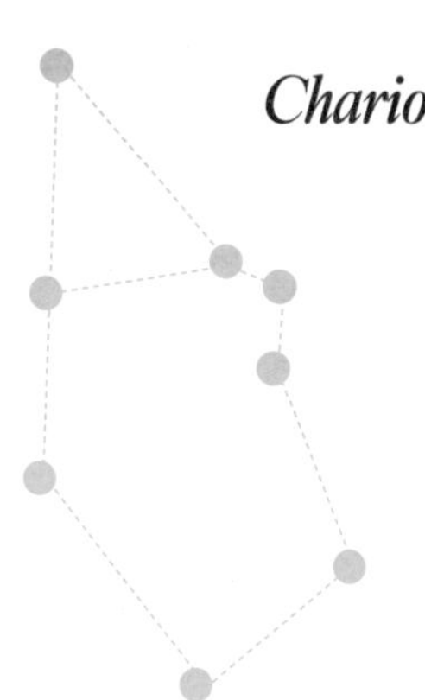

Lore

Stars within Auriga are believed to have first been recorded in Mesopotamia as an ancient constellation called GAM, which represented a scimitar or crook. In Greek mythology, Auriga often represents the hero Erichthonius of Athens, the son of Hephaestus who was born from the earth. Erichthonius was raised by the goddess Athena, and was said to be the inventor of the four-horse chariot.

How to see it

Auriga is a northern constellation, visible at latitudes between +90° and -40° and best viewed in February. It sits above the head and arm of Orion (p. 86) and next to the twins of Gemini (p. 68). It can be found by identifying the bright red star Betelgeuse, which features in both Orion and Gemini. Next, locate Pleiades in Taurus (p. 104). Draw lines from each star to the north – Auriga sits where the lines cross.

Deep-sky objects

Auriga contains the galactic anticenter, which means it is exactly opposite to the galactic centre. There are fewer stars in this region of the sky from our own galaxy, and the stars that can be seen are at the edge of the Milky Way. This area of the sky contains several open clusters, including NGC 1960 and M36. Two meteor showers can be seen in this direction – Aurigids in September and Zeta Aurigids in December/January.

Brightest star

Alpha Aurigae, or Capella, is the sixth-brightest star in the night sky. It is a quadruple star system made up of two binary pairs.

Area and main stars

657 square degrees with 8 main stars

BOÖTES

The Herdsman

Lore

Boötes is said to have been placed in the sky by Dionysus to commemorate the tragic story of Icarius. Dionysus taught Icarius how to make wine from grapes. After a night of drinking his wine, some shepherds awoke with hangovers. They thought Icarius had tried to poison them and, in a rage, they killed him.

How to see it

Boötes is a northern constellation, visible at latitudes between +90° and -50° and best viewed in June. It only has two bright stars, but they are both distinctive and easy to spot. The bright red star Arcturus is the fourth-brightest star in the sky. The other bright star is Izar. Arctarus and Izar are in the lower part of the constellation. The body of the herdsman is formed by six stars arranged in a diamond or kite shape. Arcturus is at the base of the kite.

Deep-sky objects

There are very few objects in this region that can be seen without large telescopes. One reason for this is the Boötes Void, which is an area of the sky that is devoid of galaxies and over 250 million light-years across.

Brightest star

Although Arcturus is about the same mass as the Sun, its radius is 25 times larger and it is 170 times more luminous. It is culturally significant across the history of human civilisation, from Ancient Mesopotamians to Aboriginal and Torres Strait Island Peoples. The Wotjobaluk people of south-east Australia know Marpeankurrk (Arcturus) as the mother of Djuit (Antares). Its appearance in the sky signals the arrival of the wood ant larvae, an important food source.

Area and main stars

907 square degrees with 7 main stars

CANCER

The Crab

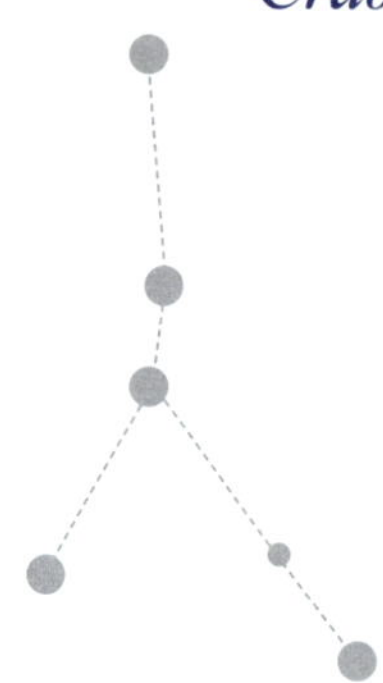

Lore

In Greek mythology, this constellation represents the crab that appeared while Heracles was fighting the serpent Hydra. The goddess Hera sent the crab to distract Heracles and help Hydra defeat him. But Heracles's strength was no match for it – he killed the crab with his foot. As a reward for the crab's service, Hera placed its image in the sky.

How to see it

Cancer is an equatorial constellation, visible at latitudes between +90° and -60° and best viewed in March. It is so faint that many people may never have seen it. Find the two brightest stars in Gemini (p. 68). Next, find Regulus, the bright star at the bottom of Leo (p. 74). Draw a line between these two locations and look for a slightly fuzzy spot. This is the heart of Cancer. The four stars around the fuzzy spot make up the body of the crab.

Deep-sky objects

At the heart of Cancer is the M44 star cluster, which can be seen well with binoculars. Another object worth looking for is M67, another open cluster that has thousands of brilliant stars.

Brightest star

Beta Cancri is an orange–red giant with a radius 50 times that of the Sun. It has one known planet, a gas giant that is 7.8 times larger than Jupiter.

Area and main stars

506 square degrees with 5 main stars

CANIS MAJOR

The Greater Dog

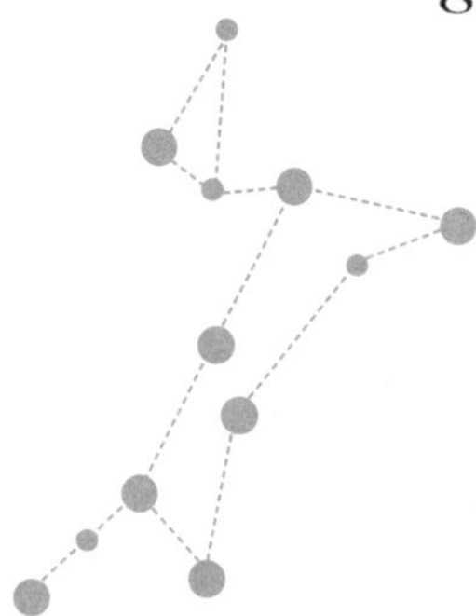

Lore

In Greek mythology, Canis Major is associated with Laelaps, who was the fastest dog in the world and destined to always catch his prey. Laelaps was given to Europa as a present from Zeus. When Cephalus took Laelaps to hunt a fox that could never be caught, Zeus realised the two animals would run for eternity. He turned them both to stone and placed Laelaps in the sky as Canis Major.

How to see it

Canis Major is a southern constellation, visible at latitudes between +60° and -90° and best viewed in February. It is one of the easiest constellations to spot. It contains Sirius, the brightest star in the night sky, which represents the dog's neck. Sirius can be found by locating Orion (p. 86) and following the line of Orion's belt to the left. The two brightest stars nearby represent the back end of the dog.

Deep-sky objects

The visible galaxies are NGC 2207 and IC 2163, along with the nebula NGC 2359.

Brightest star

Sirius is not only the brightest star in Canis Major, but the entire sky. It is the fifth closest star to Earth. Sirius is a binary star system with a main sequence star that is twice the mass of the Sun. The other star is a white dwarf that is smaller than Earth. Sirius is important to many cultures. Polynesian people use it as a navigational marker, as it matches the latitude of the archipelago of Fiji.

Area and main stars

380 square degrees with 8 main stars

CANIS MINOR

The Lesser Dog

Lore

Canis Minor is often associated with one of the two hunting dogs of Orion, but it can also be traced back to the mythology of Boötes. Canis Minor was the dog of Icarius, who was killed by shepherds who thought he gave them poisoned wine. In Ancient Egypt, this area of the sky was said to represent Anubis, the Jackal God.

How to see it

Canis Minor is a northern constellation, visible at latitudes between +90° and -75° and best viewed in February. The brightest star of Canis Minor is Procyon, which is easy to find. First, locate the bright red supergiant Betelgeuse (p. 68 and p. 86) and the brightest star in the sky, Sirius (p. 36). Procyon is nearby and forms a triangle with these two stars. The shape of the dog can be made with Procyon and the next brightest star, Gomeisa.

Deep-sky objects

Given the position of Canis Minor in the Milky Way, the deep-sky objects in this region, which include NGC 2459 and NGC 2394, can be hard to see without large telescopes. One event that stargazers can easily enjoy, however, is a meteor shower that radiates from this region of the sky, peaking on 10 and 11 December.

Brightest star

Procyon is a binary star system and the eighth-brightest star in the night sky. The two stars of Procyon are an interesting pair. The first is a main sequence star that is actively burning hydrogen in its core, like the Sun. The second is a white dwarf that is the dead remnant of a star similar to its companion.

Area and main stars

183 square degrees with 2 main stars

CAPRICORNUS

The Sea Goat

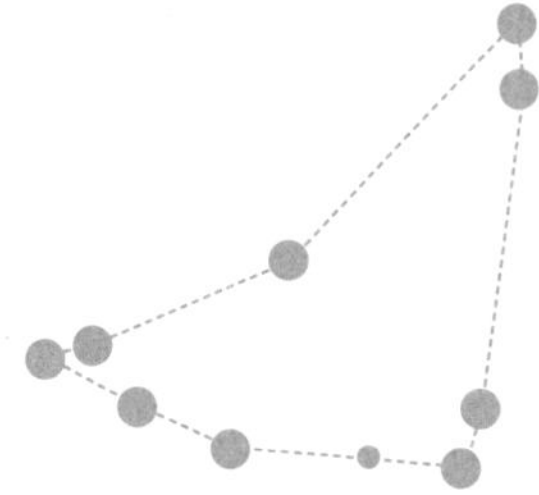

Lore

Capricornus was first recorded in the Babylonian star catalogue, which dates back to 1000 BCE. The half-goat, half-fish figure can be traced to Sumerian symbolism and Babylonian, Syrian and Egyptian mythologies. These mythological associations are among the night sky's oldest. There are multiple fables about the creation of Capricornus. One attributed to Ancient Greece tells of the god Pan, who saved himself from Typhon, a monstrous, serpentine storm giant, by transforming the lower part of his body into that of a fish and diving into a river.

How to see it

Capricornus is an equatorial constellation, visible at latitudes between +60° and -90° and best viewed in September. Its brightest star forms the top-left corner of its upside-down triangular shape.

Deep-sky objects

One object that stands out is the large open cluster M30. M30 appears to sparkle when viewed with binoculars or telescopes because of the light from several hundred thousand stars.

Brightest star

Delta Capricorni is a binary star system comprising Delta Capricorni Aa and Delta Capricorni Ab. Aa is twice the mass of the Sun, while Ab is only 0.73 times the mass of the Sun. The two stars orbit each other every 24.5 hours.

Area and main stars

414 square degrees with 9 main stars

CASSIOPEIA

The Seated Queen

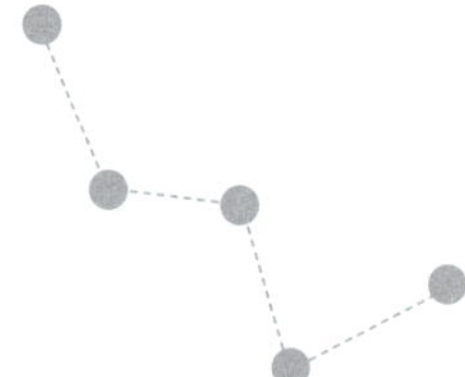

Lore

In Greek mythology, Cassiopeia was the wife of King Cepheus of Ethiopia. She was known for boasting that she and her daughter, Andromeda, were more beautiful than all the sea nymphs. After Andromeda and Perseus were wed, a former suitor claimed that he had the right to marry her. A fight broke out and Perseus used the head of Medusa to defeat his opponents, but in the chaos, he killed the king and queen. Poseidon placed Cassiopeia in the sky, doomed to circle the celestial pole forever and spend half the year upside down as punishment for her vanity.

How to see it

Cassiopeia is a northern constellation, visible at latitudes between +90° and -20° and best viewed in November. It can be found beneath the tail of Ursa Minor (p. 110) and identified by the W-shape formed by five bright stars.

Deep-sky objects

Cassiopeia contains the open clusters M52 and M103, the Heart Nebula (IC 1805a) and the Soul Nebula (IC 1848). The Heart Nebula is one of the most interestingly shaped nebulae – it looks similar to a human heart.

Brightest star

Alpha Cassiopeiae, or Schedar, is a red giant. As it nears the end of its life, it is fusing helium in its core. Along with four other stars, it forms the Legs mansion (奎宿) – one of the twenty-eight mansions of the Chinese constellations.

Area and main stars

598 square degrees with 5 main stars

CENTAURUS

The Centaur

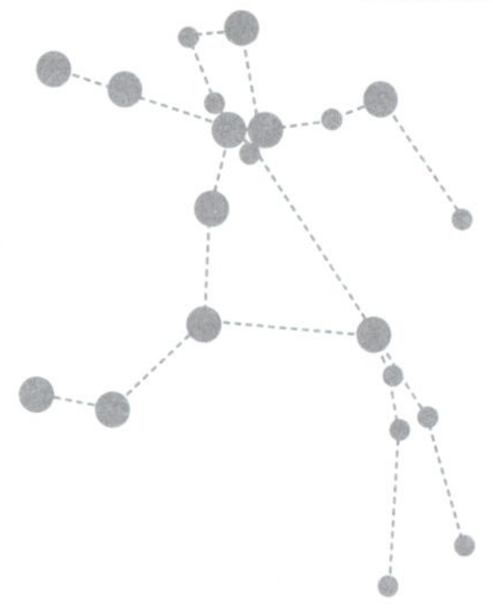

Lore

In Greek mythology, when Pholus was entertaining Heracles in his cave, he offered him wine that Dionysus had left him. The smell of the wine enticed other centaurs into the cave, and Heracles fired his poisoned arrows at them. Pholus picked up an arrow to examine it, dropped it on his foot and died.

How to see it

Centaurus is a southern constellation, visible at latitudes between +25° and -90° and best viewed in May. The two bright stars that form the centaur's front feet point towards Crux (p. 136) (the Southern Cross), which sits below Centaurus.

Deep-sky objects

In this constellation, a bright globular cluster of over 10 million stars is large enough to be seen with the naked eye. This cluster is Centaurus A. It is the fifth-brightest galaxy in our skies, making it easy to see through telescopes.

Brightest star

Alpha Centauri is a triple star system that can be seen with the naked eye. It has two Sun-like stars, Alpha Centauri A and B, and a small red dwarf, Alpha Centauri C. At just 4.24 light-years away, Alpha Centauri C is the closest star to our solar system. The Boorong people of north-western Victoria, Australia, know Alpha Centauri and Beta Centauri (the second-brightest star in Centaurus) as Berm Berm-gle, the brothers who killed Tchingal, the celestial emu. In Indigenous Australian astronomy, the celestial emu is a dark constellation made up of dust clouds in the Milky Way.

Area and main stars

1060 square degrees with 11 main stars

CEPHEUS

The King

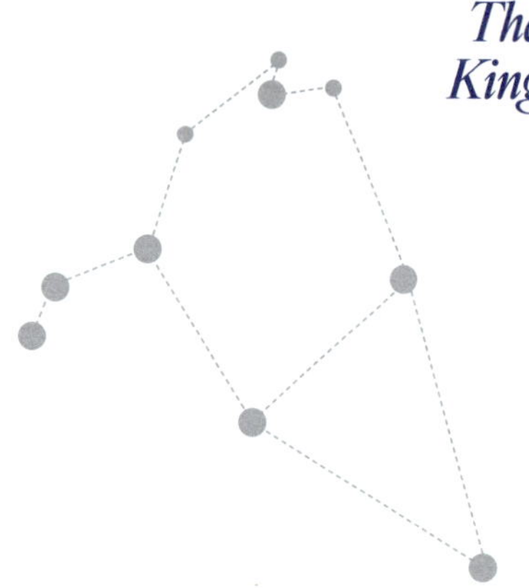

Lore

In Greek mythology, Cepheus was the son of Io, the husband of Cassiopeia and the father of Andromeda. Cepheus met his end at his daughter's wedding to Perseus. During the festivities, Cepheus's brother, Phineas, appeared and claimed he had been promised Andromeda first. A fight broke out and Perseus used Medusa's head to turn his enemies into stone. In the chaos, he accidentally turned Cepheus and Cassiopeia into stone as well.

How to see it

Cepheus is a northern constellation, visible at latitudes between +90° and -10° and best viewed in November. Start by locating Polaris (the North Star), in Ursa Minor (p. 110). Next, find the next brightest star near Polaris. This is Errai. The body of Cepheus is made up by a square joined to a triangle. Errai is the star at the point of the triangle, and represents Cepheus's left foot.

Deep-sky objects

In Cepheus, both the Fireworks Galaxy (NGC 6946) and the open cluster NGC 188 are visible.

Brightest star

Alpha Cephei, or Alderamin, is 820 million years old and twice the mass of the Sun. It is located near the path traced in the sky by an imaginary line extending from Earth's North Pole. Alpha Cephei will eventually become the North Star. This is expected expected to occur around the year 7500 CE.

Area and main stars

588 square degrees with 7 main stars

CETUS

The Sea Monster

Lore

In Greek mythology, Cetus was often associated with a sea monster ruled by Poseidon. This monster that was sent by Poseidon to attack the home of Cassiopeia. It was slain by Perseus when it was turned to stone using Medusa's head.

How to see it

Cetus is an equatorial constellation, visible at latitudes between +70° and -90° and best viewed in November. It is easiest to find in dark skies and with no objects obstructing the horizon. From the top of Pisces (p. 92), look down slightly to find stars that appear to form a circle. This is the head of Cetus.

Deep-sky objects

Viewers can spot M77, a face-on spiral galaxy, with a small telescope. Several other fainter galaxies, such as NGC 247 and NGC 17, can be seen using larger telescopes.

Brightest star

Beta Ceti is a K-type orange giant. It is 3.5 times the mass of the Sun and has been observed to flare for long durations. These flaring outbursts can increase the star's brightness for days and have even been detected by X-ray telescopes in space.

Area and main stars

1231 square degrees with 14 main stars

CORONA AUSTRALIS

The Southern Crown

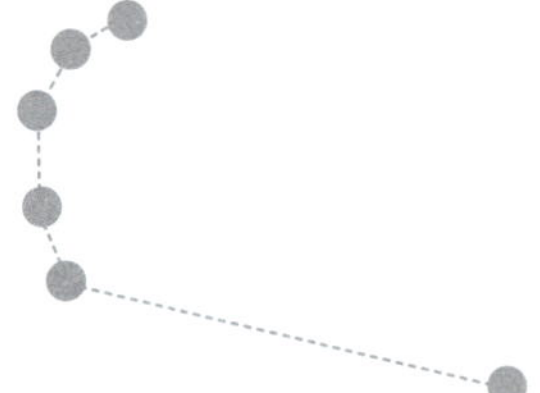

Lore

○ Corona Australis is sometimes associated with the Greek god Dionysus. The stars represent the crown that Dionysus placed in the sky after freeing his mother, Semele, from Hades. Corona Australis has represented many objects throughout history, such as the Southern Wreath, the Little Crown and the Southern Coil. Arabian astronomers also saw it as a tortoise and an ostrich's nest.

How to see it

✦ Corona Australis is a southern constellation, visible at latitudes between +40° and -90° and best viewed in August. Locate Scorpius (p. 100). Trace its stars, then go outwards from the bottom of the scorpion's tail, just where it starts to bend. Corona Australis sits here. It appears as a small circle that can be made by joining the faint stars.

Deep-sky objects

⊙ A large telescope is needed to see the deep-sky objects in this region of the sky. Viewers can look for the Corona Australis Nebula, the Coronet Cluster and the nebula NGC 6729.

Brightest star

⁂ Alpha Coronae Australis is just 125 light-years from Earth. It is an A-type star, actively burning hydrogen in its core.

Area and main stars

◗ 128 square degrees with 6 main stars

CORONA BOREALIS

The Northern Crown

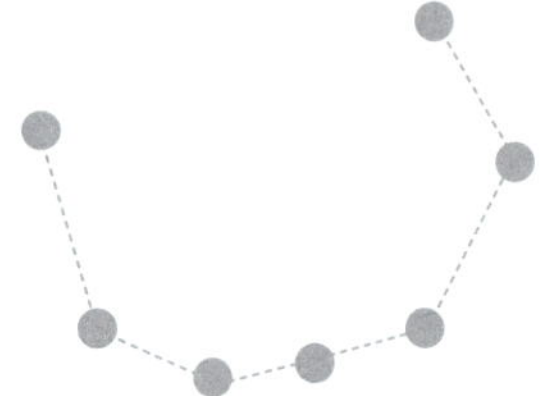

Lore

○ Corona Borealis is often associated with the Greek myth of Ariadne, a princess of Crete. Ariadne is most famous for helping Theseus defeat the Minotaur, which was locked inside a labyrinth. The couple sailed to the island of Naxos, where Theseus abandoned Ariadne. The god Dionysus found the princess weeping alone and fell in love with her. When they were married, Ariadne wore a crown made by Hephaestus. She threw it into the sky and the gems became stars.

How to see it

✦ Corona Borealis is a northern constellation, visible at latitudes between +90° and -50° and best viewed in June. Find the bright red star Arcturus in Boötes (p. 32). To the left of Boötes, you will see three bright stars in a line that form part of the crown. The full crown is as a U-shape made up of seven stars sitting closely together. The brightest of these stars, known as Gemma or Alphecca, is in the middle.

Deep-sky objects

⊙ Two notable galaxies are visible in Corona Borealis: the spiral galaxy NGC 6085 and the elliptical galaxy NGC 6086.

Brightest star

⁂ Alpha Coronae Borealis is a binary star system that is visible to the naked eye. The two stars are 2.58 and 0.92 times the mass of the Sun. The larger star is suspected to still have a large disk of dust around it, based on observations made using infrared radiation.

Area and main stars

◗ 179 square degrees with 7 main stars

CORVUS

The Crow

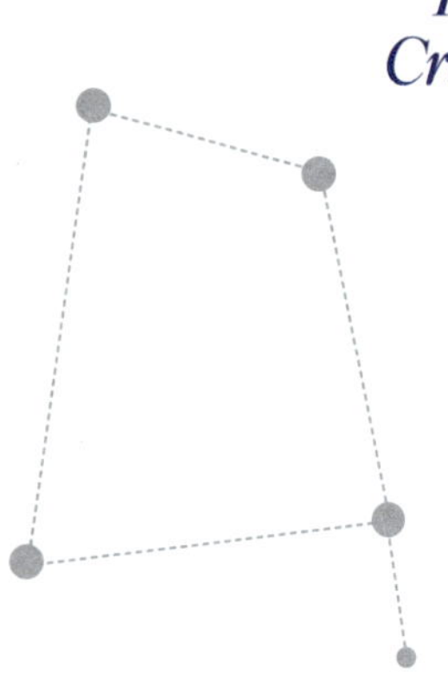

Lore

Corvus is one of the forty-eight constellations catalogued by the Greek astronomer Ptolemy (100–170). Its stars are found in Babylonian star catalogues made in 1100 BCE. Corvus, Crater, Aquila and Piscis were introduced by the Greeks around 500 BCE and marked the winter and summer solstices. Corvus is linked to the myth of Apollo. When a white crow told Apollo that his lover Coronis was unfaithful, the god turned its feathers black in a fit of rage.

How to see it

Corvus is a southern constellation, visible at latitudes between +60° and -90° and best viewed in May. First, look south and find Spica, the brightest star in Virgo (p. 112). Corvus is to the west of Spica. Its four main stars are the brightest in that region of the sky.

Deep-sky objects

Corvus has no Messier objects but it does contain several other galaxies and NGC 4361, a planetary nebula that can be seen with home telescopes. NGC 4361 – which can easily be mistaken for a galaxy, due to its shape – sits at the centre of Corvus. A small meteor shower, Eta Corvids, occurs between 20 January and 26 January.

Brightest star

Gamma Corvi is a binary star system. Its traditional name, Gienah, is derived from Arabic and means 'the right wing of the crow'. The largest star in Gamma Corvi is 4.2 times the mass of the Sun, and the second-largest is estimated to be 0.8 times the mass of the Sun.

Area and main stars

184 square degrees with 4 main stars

CRATER

The Cup

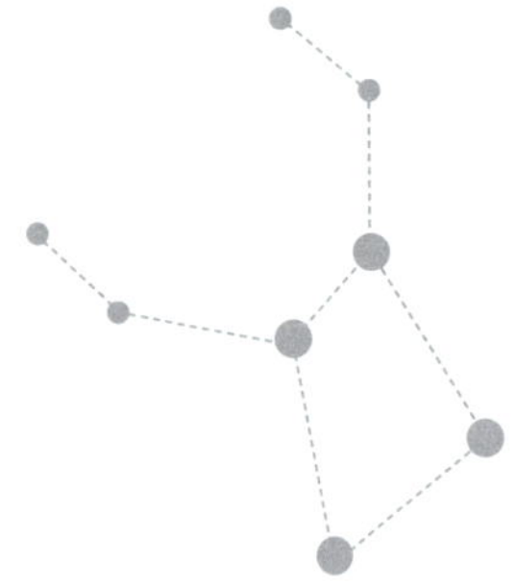

Lore

The constellation Crater is often associated with the Greek myth about Apollo and a crow. The crow was asked to fetch a cup of water for the god. As he was flying to the spring, he noticed a grove of fig trees. He landed on one and waited several days for the figs to ripen so he could eat them. When he finally got to the spring, he realised that Apollo would be angry about the delay. He spotted a snake and flew back with it in its claws as an excuse. Apollo, the god of truth, saw through this lie and cursed the crow.

How to see it

Crater is a southern constellation, visible at latitudes between +65° and -90° and best viewed in April. It is not a large constellation, but its near-symmetric shape makes it easy to spot. Look for the bright stars Spica, in Virgo (p. 112), and Regulus, in Leo (p. 74). Draw a line between them. Crater, made up of fainter stars, sits above this line, roughly near the middle.

Deep-sky objects

Only faint deep-sky objects are found in this region of the sky. These include galaxies NGC 3981 and NGC 3887, which can only be seen with large telescopes.

Brightest star

Delta Crateris is an orange giant with a radius 20 times that of the Sun. Crater itself is a faint constellation, so Delta Crateris doesn't stand out in the night sky, but it is visible with the naked eye.

Area and main stars

282 square degrees with 4 main stars

CYGNUS

The Swan

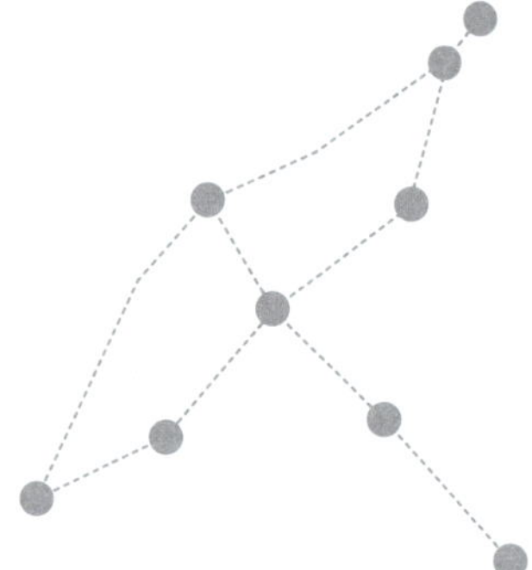

Lore

Cygnus is one of the forty-eight constellations catalogued by the Greek astronomer Ptolemy (100–170). It has been identified and used across many cultures for millennia. In Tonga it was known as Tuula-lupe, and in New Zealand it was Mara-tea. In Greek mythology, Cygnus is linked to several significant swans.

How to see it

Cygnus is a northern constellation, visible at latitudes between +90° and -40° and best viewed in September. It is very large constellation and best found by locating the cross within it. Deneb, its brightest star, forms the top of the cross.

Deep-sky objects

Cygnus is home to many interesting objects, from star clusters to black holes. Visible to the naked eye in dark clear skies is the M39 open cluster. It is 950 light-years from Earth and has thirty stars loosely arranged over a wide area. There is also NGC 6826, a planetary nebula 2000 light-years from Earth. An object that is not visible to us is Cygnus X-1, the first black hole ever discovered. It sits in the bottom half of the cross, is 7300 light-years away, and is 21 times the mass of the Sun.

Brightest star

Deneb is the nineteenth-brightest star in the sky. Its name is derived from the Arabic word for 'tail'. Deneb is a blue supergiant. It has a mass 15.5 times that of the Sun and is 55,000 times more luminous.

Area and main stars

804 square degrees with 9 main stars

DELPHINUS

The Dolphin

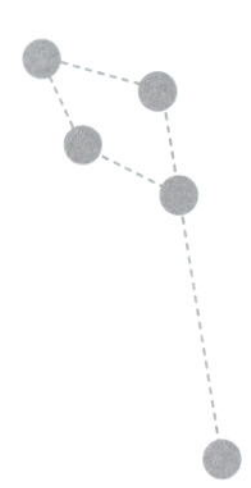

Lore

Delphinus is a small northern constellation that is typically associated with two Greek myths. The first is related to Poseidon's desire to marry Amphitrite. To protect her purity, Amphitrite ran away to the Atlas Mountains. Poseidon sent searchers to find her, including Delphinus. Delphinus convinced Amphitrite to accept Poseidon's wooing, and Poseidon placed the dolphin in the sky as a sign of his gratitude. The other myth relates to Greek poet Arion of Lesbos, who was saved by a dolphin when he jumped into the sea. In Hindu astrology, the stars in Delphinus correspond to a nakshatra, or lunar mansion, in the Northern Sky.

How to see it

Delphinus is a northern constellation, visible at latitudes between +90° and -69° and best seen in September. Although it is one of the smaller constellations, it is easy to find. Locate the bright stars Altair in Aquila (p. 24), Vega in Lyra (p. 82) and Deneb in Cygnus (p. 58). These stars form the corners of the recognisable giant summer triangle. Delphinus sits directly beneath, towards the south.

Deep-sky objects

There are many deep-sky objects in Delphinus, including two planetary nebulae. NGC 6891 has three distinct shells and was formed when a star died and expanded its layers. NGC 6905 is known as the Blue Flash Nebula, due to its colour.

Brightest star

Beta Delphini, or Rotanev, is a binary star system comprising two F-type stars. One is 1.75 times the mass of the Sun and the other is 1.47 times.

Area and main stars

189 square degrees with 5 main stars

DRACO

The Dragon

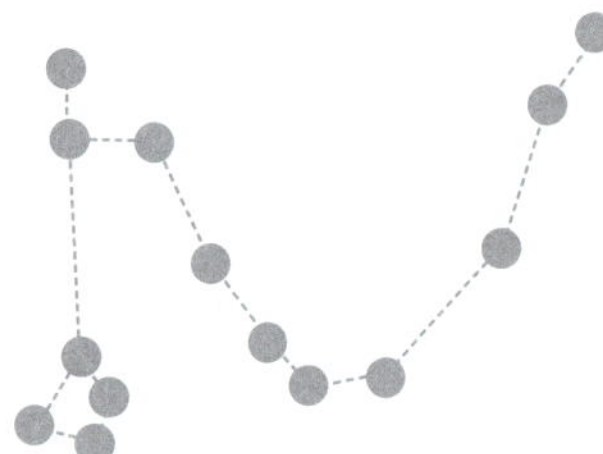

Lore

Draco has many interesting mythological stories arising from multiple cultures. One is the Babylonian creation story of Tiamat, who turned herself into a dragon but, when defeated, was split in two. One half became the stars, the other remained on Earth. In Egyptian lore, Draco was seen as a hippopotamus or crocodile, representing the different gods in those forms. In Greek mythology, one story tells of Hercules trying to obtain an apple from the golden apple tree in Hera's garden, which was protected by a dragon. Hercules succeeded in killing the dragon. Hera, saddened by this, set the dragon's image forever in the stars.

How to see it

Draco is a northern constellation, visible at latitudes between +90° and -15° and best viewed in July. One way to locate it is to find the Big Dipper in Ursa Major (p. 108) and the Little Dipper in Ursa Minor (p. 110). They are on either side of the dragon's body. Join the stars that run between them to trace the tail and head of the dragon's shape.

Deep-sky objects

Objects seen in Draco include the Cat's Eye Nebula (NGC 6543) and the galaxies NGC 4125 and NGC 5866.

Brightest star

Gamma Draconis is a 1.3-billion-year-old giant. It has a mass 2.14 times greater than the Sun and it is 598 times more luminous.

Area and main stars

1083 square degrees with 14 main stars

EQUULEUS

The Little Horse

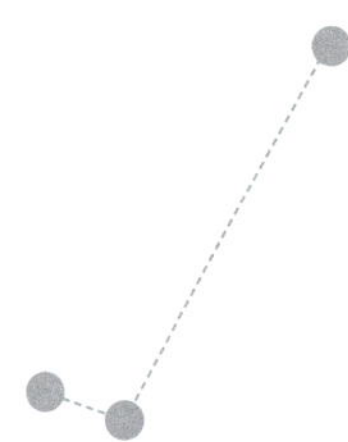

Lore

Equuleus is the second-smallest constellation in the sky. In Greek mythology, it is associated with the foal Celeris (meaning 'speed'), who was sired by Pegasus. Other myths link Equuleus to the horse who was born when Poseidon's trident struck the earth during a contest with Athena. In Chinese astronomy, the stars that make up Equuleus are located within the Black Tortoise of the North – one of the four symbols of the Chinese constellations.

How to see it

Equuleus is a northern constellation, visible at latitudes between +90° and -80°and best seen in September. As it is one of the smallest constellations in the sky, first find Pegasus (p. 88). Next, look for the bright star Altair, in Aquila (p. 24). Equuleus sits between Altair and the head of Pegasus.

Deep-sky objects

This region of the sky is very far away, so you won't see many deep-sky objects with a home telescope. Larger telescopes reveal multiple galaxies, including two spiral galaxies – NGC 7015, 203 million light-years away, and NGC 7040, 260 million light-years away. There is also a binary star system, NGC 7045, which was first mistaken for a nebula but is actually in our own galaxy.

Brightest star

Alpha Equulei, or Kitalpha, is a G-type giant. It is a spectroscopic binary with a companion A-type dwarf star. Their masses are 2.3 and 2.0 times that of the Sun, respectively, and the G-type star is twice as luminous.

Area and main stars

72 square degrees with 3 main stars

ERIDANUS

The River Eridanus

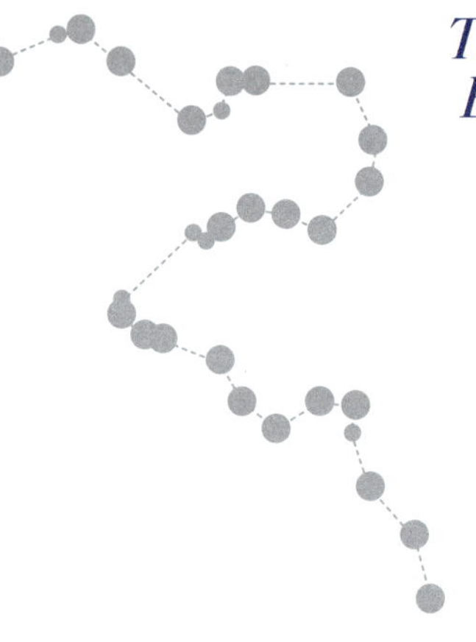

Lore

The river in the sky, known as Eridanus, is associated with the mythology of the lyre player Orpheus. Legend says that Orpheus wandered the lands playing songs in memory of his lost wife, Eurydice, until he encountered a group of worshippers of Dionysus. They tried to get Orpheus to join in their partying, but when he refused, they killed him, throwing his body and his lyre into a nearby river. Apollo placed the lyre and the river among the stars.

How to see it

Eridanus is a southern constellation, visible at latitudes between +32° and -90° and best viewed in December. It is a long constellation that travels a vast distance across the sky. Locate the belt of Orion (p. 86). Look towards his left foot at the star Cursa, which is close to Orion's brightest star, Rigel. From Cursa, the stars that form Eridanus trail across the sky, winding in a reverse S-shape towards the Small Magellanic Cloud.

Deep-sky objects

There are multiple deep-sky objects in this patch of sky. Galaxies include NGC 1531, NGC 1532 and NGC 1232. Nebulae include Cleopatra's Eye (NGC 1535) and the Witch Head Nebula (IC 2118).

Brightest star

Alpha Eridani A, or Achernar, is the ninth-brightest star in the night sky. It is part of a binary star system orbiting Alpha Eridani B. Achernar is a large blue star, approximately 6 times the mass of the Sun.

Area and main stars

1138 square degrees with 24 main stars

GEMINI

The Twins

Lore

Represented by a set of twins, often holding hands, this constellation depicts one mythology around Pollux and Castor. These inseparable half-brothers stole and wed the fiancées of their cousins Idas and Lynceus. The rivalry continued until a battle broke out. Castor was killed, and Pollux was injured but survived. Devastated, Pollux pleaded with his father, Zeus, to bring Castor back to life or to let him join Castor in the underworld. Zeus arranged a compromise – the pair could spend half their time in the underworld and half their time in the heavens.

How to see it

Gemini is an equatorial constellation, visible at latitudes between +90° and -60° and best viewed in February. Find the belt of Orion (p. 86) and, above that, locate the bright red star of Betelgeuse. From here, look east for two bright stars next to each other. These are the heads of the twins, Pollux and Castor.

Deep-sky objects

Two stunning planetary nebulae, NGC 2392 and the Medusa Nebula (Abell 21), can be seen in Gemini.

Brightest star

Pollux (Beta Geminorum) is a red giant. It is only 34 light-years from Earth, making it the closest red giant to us. In 2006, an exoplanet was discovered orbiting Pollux. This exoplanet is estimated to be twice the mass of Jupiter.

Area and main stars

514 square degrees with 17 main stars

HERCULES

The Heracles

Lore

The Hercules constellation represents the Roman version of the Greek hero Heracles. In Greek mythology, Heracles was the son of Zeus and a mortal woman, Alcmene. After being driven mad by Hera and murdering his own wife and children, Heracles was sent to serve his archenemy, Eurystheus, the king of Tiryns and Mycenae. For the next twelve years, Heracles completed several tasks – known as the Labours of Heracles – doing everything from fighting monsters to visiting the underworld. Zeus eventually placed him in the sky.

How to see it

Hercules is a northern constellation, visible at latitudes between +90° and −50° and best viewed in July. First, find the stars Vega in Lyra (p. 82) and Arcturus in Boötes (p. 32). Draw a line connecting them. One-third of the way along this line, four bright stars make a square. This square is the lower torso of Hercules. From here, trace the legs towards the north and connect the rest of the body and the arms.

Deep-sky objects

Two globular clusters are located in Hercules: M13 and M92. Both can be seen with binoculars or small telescopes.

Brightest star

Beta Herculis is a binary star system that is estimated to be 420 million years old. The two stars are 2.91 and 0.9 times the mass of the Sun.

Area and main stars

1225 square degrees with 14 main stars

HYDRA

The Water Snake

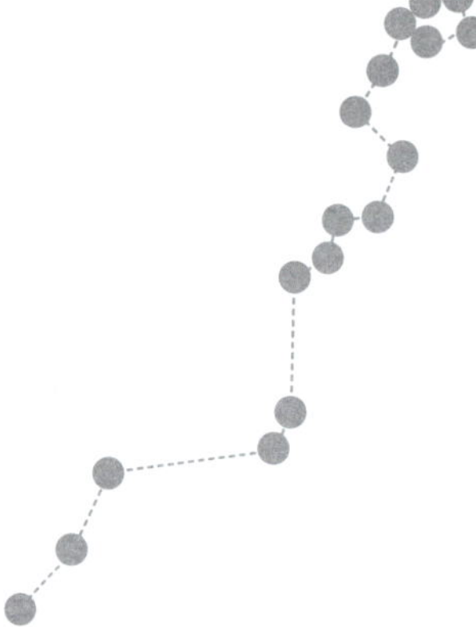

Lore

○ The Greek myth of Hydra is an adaption of a Babylonian constellation that referenced a snake. For the Ancient Greeks, Hydra was the water snake that a crow served to Apollo, who then cast it into the sky. Another Greek myth relates to a sea serpent that could regrow two heads if beheaded. In Chinese astronomy, the stars of Hydra sit in the Vermilion Bird and the Azure Dragon, two of the four symbols in Chinese constellations.

How to see it

✦ Hydra is one of the largest constellations in the night sky. It is a southern constellation, visible at latitudes between +54° and -83° and is best viewed in April. The head of the snake is located south of Cancer (p. 34) and the tail sits between Centaurus (p. 44) and Libra (p. 78).

Deep-sky objects

⊙ There are three Messier objects in Hydra. One is the Southern Pinwheel Galaxy (M83) is 15 million light-years away. It is one of the closest and brightest galaxies of its type in the night sky, and is a wonderful object to target with binoculars or small telescopes. Another is NGC 3242, a planetary nebula 1400 light years from Earth. It is nicknamed Ghost of Jupiter and appears blue in small telescopes.

Brightest star

⁂ Alpha Hydrae, or Alphard, is a K-type giant that is 3.2 times the mass of the Sun and sits 177 light-years from Earth. It's a well-known star, and appears on the flag of Brazil.

Area and main stars

◗ 1303 square degrees with 17 main stars

LEO

The Lion

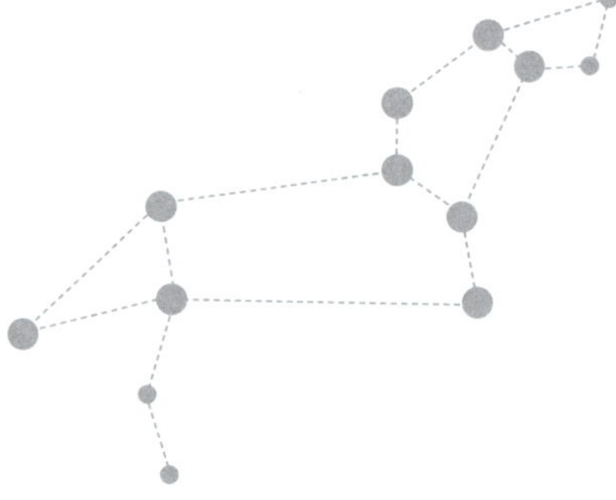

Lore

In Greek mythology, this constellation represents an epic battle between a lion and the great hero Heracles, the son of Zeus and Alcmene. The legend goes that Heracles was tasked by his cousin Eurystheus to slay a man-eating lion that was terrorising the countryside. He spotted the lion but his arrows were not effective, so Heracles followed the lion back to its den. He sealed one of the entrances to the den with a rolling stone and entered the other, eventually succeeding in slaying the lion. Zeus put the lion in the sky to honour his son's victory.

How to see it

Leo is an equatorial constellation, visible at latitudes between +90° and -65° and best viewed in April. The easiest way to locate it is to look for a back-to-front question mark, or sickle-like shape, made from the brightest stars in the constellation. This is the head of the lion.

Deep-sky objects

There are many unique galaxies in Leo. Three notable galaxies near the back legs of the lion shape are M66, M65 and NGC 3628.

Brightest star

The brightest star in Leo, Regulus, represents the Lion's heart. Regulus appears as a single star, but is actually a quadruple star system, composed of two pairs of binary stars. The largest star in this system is 3.4 times the mass of the Sun and 314 times more luminous than the Sun too.

Area and main stars

947 square degrees with 9 main stars

LEPUS

The Hare

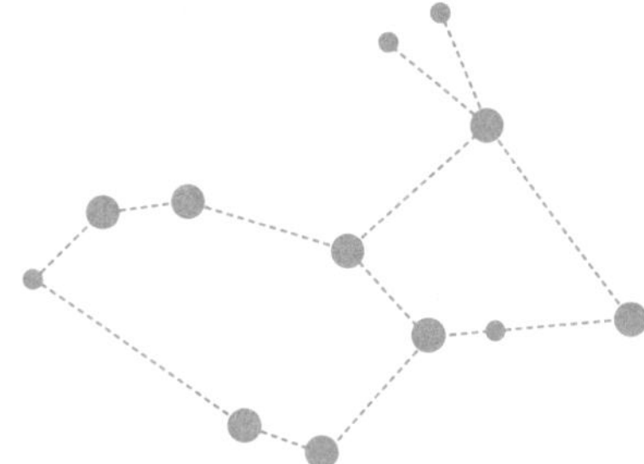

Lore

Lepus is not associated with any particular myth, although it is sometimes said that it represents a hare being pursued by Orion's hunting dogs. It has also been associated with the Moon Rabbit, which is a mythical figure in East Asian and Indigenous American folklore. Its shape mimics the dark seas of the Moon.

How to see it

✦ Lepus is a southern constellation, visible at latitudes between +63° and -90° and best seen in January. Although a small constellation, it is an easy one to spot. It sits in between two bright stars: Sirius in Canis Major (p. 36), and Rigel in Orion (p. 86).

Deep-sky objects

There is one Messier object in Lepus. M79 is a globular cluster located 42,000 light-years from Earth.

Brightest star

Alpha Leporis, also referred to as Arneb, is a yellow supergiant with a mass 14 times greater than the Sun. It is 2200 light-years from Earth and estimated to be only 13 million years old. It is the central star of this constellation.

Area and main stars

290 square degrees with 8 main stars

LIBRA

The Balance

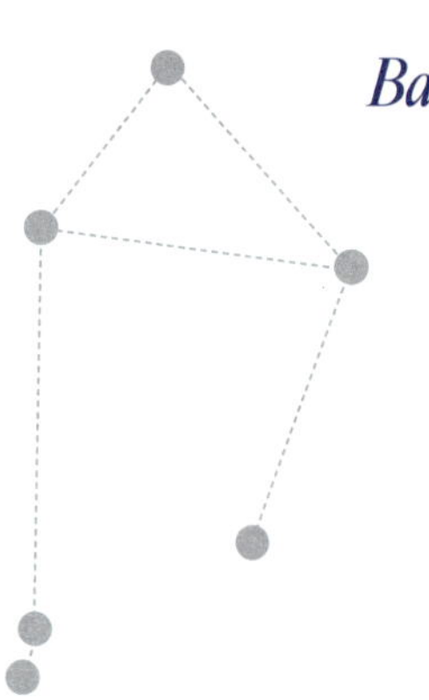

Lore

○ Libra was often depicted as a god or goddess holding a pair of scales. This imagery represents the mythology around the goddess Astraea's observations of humans after the gods left Earth. Astraea weighed the good in human souls against the bad. Arabian astronomers commonly represented this constellation as scales and this is how it remains today.

How to see it

✦ Libra is a southern constellation, visible at latitudes between +65° and -90° and best viewed in June. Find the bright stars Antares in Scorpius (p. 100) and Spica in Virgo (p. 112). Draw a line between them. The top of the scales is roughly halfway along the line. The three brightest stars in this constellation form a triangle.

Deep-sky objects

⊙ Although not classed as a deep-sky object, the star Methuselah is remarkable and worth trying to spot. Methuselah is believed to be the oldest known star in the universe, formed shortly after the Big Bang, 13.8 billion years ago.

Brightest star

⁂ Beta Librae is a B-type main sequence star that is 130 times more luminous than the Sun. It is 185 light-years from Earth. Beta Librae is expected to run out of fuel in its core in 200 million years.

Area and main stars

◗ 538 square degrees with 6 main stars

LUPUS

The Wolf

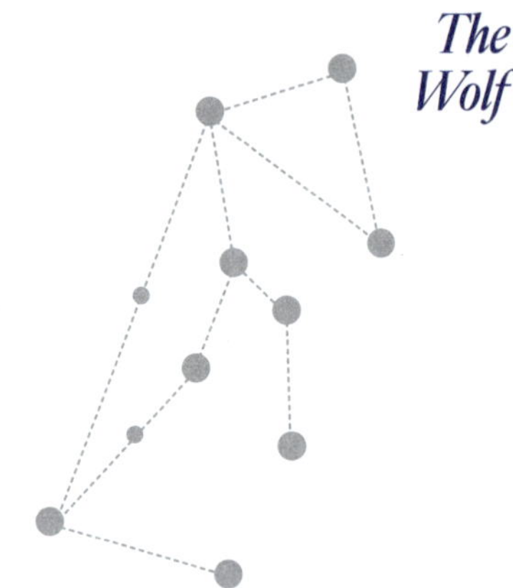

Lore

Although Lupus is one of the forty-eight Ptolemaic constellations, it isn't associated with classical mythology. The stars that form Lupus once belonged to the constellation Centaurus. The symbolism of a wolf or dog can be traced back to Babylonian lore about Uridimmu (Mad Dog), a mythical lion-like creature with a human head.

How to see it

Lupus is a southern constellation, visible at latitudes between +35° and -90° and best viewed in June. Find the red star Antares at the heart of Scorpius (p. 100). Next look for the two bright stars that make up the front legs of Centaurus (p. 44). Draw a line between these two locations, then look slightly above it. Lupus travels almost the length of this line. The head of the wolf is closest to Antares.

Deep-sky objects

Use binoculars to spot the NGC 5986 globular cluster. Also within Lupus, but too faint to see without large telescopes, are the planetary nebulae IC 4406 and the supernova remnant SN 1006. SN 1006 is possibly the brightest stellar event ever observed.

Brightest star

Alpha Lupi is a blue giant. Its mass is 10 times that of the Sun, and it is a candidate for a future near-Earth supernova. Its age is estimated to be less than 20 million years, so it is likely to explode in the next 5 million years.

Area and main stars

334 square degrees with 9 main stars

LYRA

The Lyre

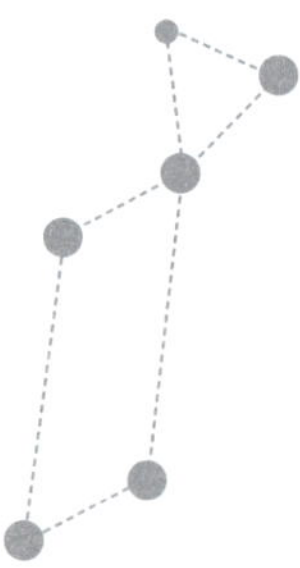

Lore

In Greek mythology, Lyra represented the lyre of Orpheus. Orpheus was said to play music that could charm inanimate objects. Eurydice, his wife, was bitten and killed by a snake. To bring her back from the dead, Orpheus ventured to the underworld and used his lyre to charm Hades, the god of the underworld.

How to see it

Lyra is a northern constellation, visible at latitudes between +90° and -40° and best viewed in August. It is one of the easier constellations to locate, thanks to its brightest star, Vega. Viewers can also first locate Cygnus (p. 58) and then look towards the west.

Deep-sky objects

One of the most popular deep-sky objects in Lyra is M57, known as the Ring Nebula, located 2550 light-years from Earth. It is estimated to be between 6000 and 8000 years old, meaning that it potentially didn't exist at the time humans invented the wheel. Another object is NGC 6745, an irregular galaxy that is actually a trio of galaxies that have been colliding with each other for millions of years.

Brightest star

Vega, or Alpha Lyrae, is the fifth-brightest star in the Northern Sky. Vega has been described by astronomers as 'arguably the next most important star in the sky, after the Sun'. This is because it has been used as a baseline to calibrate our brightness measurements of stars. Vega is an A-type star that is a little over 2 times the mass of the Sun, and only one-tenth the age.

Area and main stars

286 square degrees with 5 main stars

OPHIUCHUS

The Serpent-Bearer

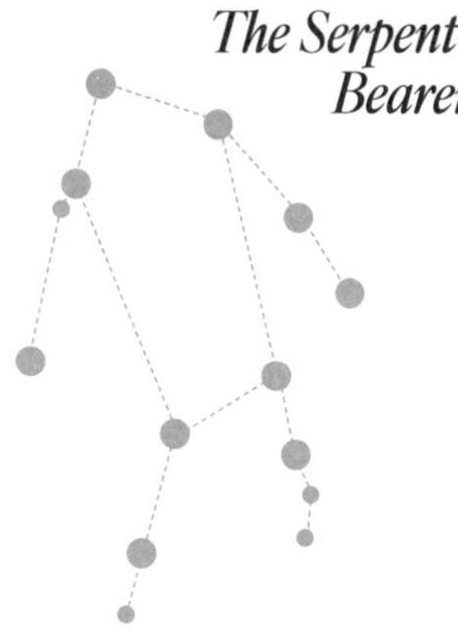

Lore

The stars of Ophiuchus represent the healer Asclepius. It was originally part of a larger constellation that included Serpens. The Greek myth tells of Asclepius, the son of Apollo, who accidentally killed a snake. After witnessing the snake come back to life when another snake placed a herb on its head, Asclepius used the herb to bring back many slain heroes. Hades was displeased by this, especially after Asclepius tried to resurrect Orion. Asclepius was killed by order of the gods and placed in the sky.

How to see it

Ophiuchus is an equatorial constellation, visible at latitudes between +80° and -80° and best viewed in July. It is not very bright and can be hard to find. Find the bright star Antares in Scorpius (p. 100). Next, look north for the star that makes up the left knee of Ophiuchus. For this constellation, it's best to use a star chart or a night-sky application to help you find the figure of Ophiuchus.

Deep-sky objects

Ophiuchus is home to Kepler's Supernova (SN 1604), M9, M10, M12 and M14. Most of these can be seen with binoculars.

Brightest star

Alpha Ophiuchi, or Rasalhague, is a binary star system. The two stars, Alpha Ophiuchi A and Alpha Ophiuchi B, orbit each other every 8.62 years. They are 2.4 and 0.85 times the mass of the Sun, respectively.

Area and main stars

948 square degrees with 10 main stars

ORION

The Hunter

Lore

This constellation is named after a hunter in Greek mythology. Orion is a symbol of strength and skill who is known for his ability to catch any animal he sets his sights on – at least, all but one. In one version of the story, Scorpius was sent to defeat Orion. After a fierce battle, Orion was stung and killed by the mighty scorpion.

How to see it

Orion is an equatorial constellation, visible at latitudes between +85° and -75° and best viewed in January. First find the belt – three bright stars that appear in a straight line. Then look nearby for a bright red star, Betelgeuse, which forms Orion's shoulder. To one side of the belt is Rigel, a bright blue star that marks the hunter's right foot.

Deep-sky objects

Orion is home to many deep-sky objects. The most notable is the Orion Nebula – the birthplace of new stars. It is one of the brightest nebulae in the night sky and is only 1344 light-years away. Another deep-sky object worth a look is the Horsehead Nebula (Barnard 33).

Brightest star

Rigel is a blue supergiant. It is 21 times the mass of the Sun and only 8 million years old. Rigel is the brightest star in a star system comprised of at least four stars in orbit around each other. To the naked eye, they all appear as a single point.

Area and main stars

594 square degrees with 7 main stars

PEGASUS

The Winged Horse

Lore

This ancient constellation represents the story of Andromeda and Perseus in Greek mythology. The legend goes that Pegasus was born after his father, Perseus, cut off the head of Medusa and some of her blood fell into the sea. Pegasus assisted a man named Bellerophon on several dangerous missions. However, Bellerophon grew arrogant and demanded to meet with the gods on Mount Olympus. Zeus, greatly annoyed by Bellerophon, sent an insect to bite Pegasus on the behind, causing him to buck Bellerophon off his back. Pegasus became beloved on Mount Olympus and carried Zeus's thunderbolts.

How to see it

Pegasus is a northern constellation, visible at latitudes between +90° and -60° and best viewed in October. The centre of Pegasus can be found by drawing a box using the four stars above Pisces (p. 92). A star chart or night-sky application will be helpful.

Deep-sky objects

The globular cluster M15 has a bright core of many brilliant stars.

Brightest star

Epsilon Pegasi is a red supergiant with a mass 12 times that of the Sun that is fusing helium in its core. It will eventually die and form a white dwarf, although it might be just large enough to form a supernova.

Area and main stars

1121 square degrees with 9 main stars

PERSEUS

The Hero

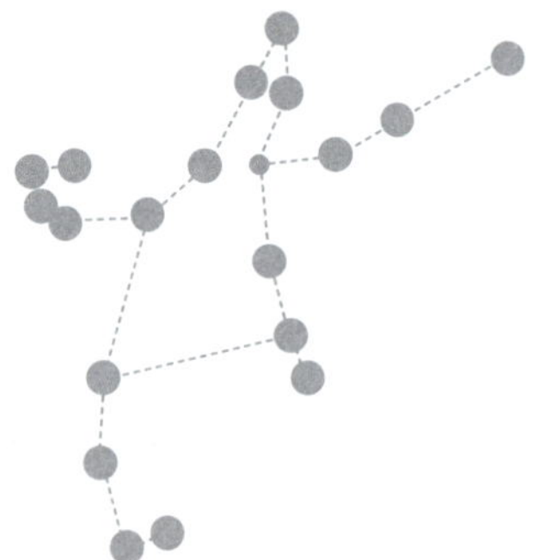

Lore

Perseus can be traced back to Greek mythology. In one myth, Perseus was the son of Danaë, who was sent by a king to return the head of Medusa. It's said that Perseus slew Medusa in her sleep. In another myth, Perseus goes on to save Andromeda, the daughter of Cepheus, from Cetus the sea monster.

How to see it

Perseus is a northern constellation, visible at latitudes between +90° and -35° and best viewed in December. It is a larger constellation that can found by locating Pleiades in Taurus (p. 104). Perseus is next to this constellation – the head of Perseus is about the same distance from Pleiades as Orion's belt.

Deep-sky objects

Because the galactic plane passes through Perseus, this area of the sky is fantastic for stargazing. Two open clusters, NGC 869 and NGC 884, are visible through binoculars or small telescopes. Viewers can also spot M76, a planetary nebula, known as the Little Dumbbell Nebula.

Brightest star

Alpha Persei, or Mirfak, is an F-type giant that is 7.3 times larger than the Sun. To Native Hawaiians, it is known as Hinali'i. The name commemorates a huge tsunami that marked the beginning of the migration of Maui.

Area and main stars

615 square degrees with 19 main stars

PISCES

The Fishes

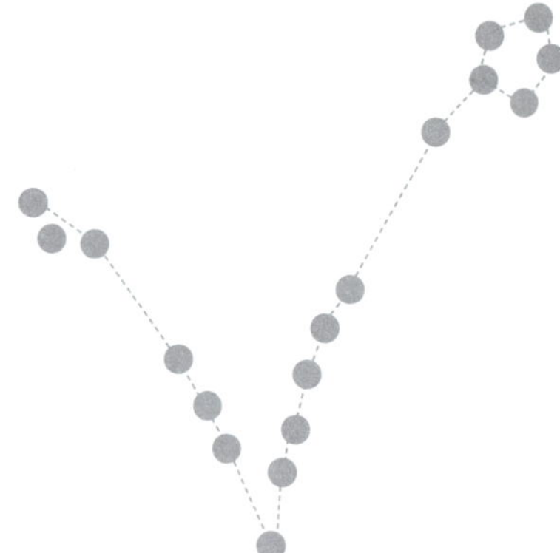

Lore

The two fish of the constellation Pisces, often drawn with a ribbon connecting them, can be traced back to both Roman and Greek mythology. Venus and Cupid (to the Romans), or Aphrodite and Eros (to the Greeks), transformed themselves into fish to escape the monster Typhon. This is a similar story to that of the constellation Capricornus.

How to see it

Pisces is an equatorial constellation, visible at latitudes between +90° and -65° and best viewed in November. One way to find it is to look for a circle or diamond shape formed by six faint stars. Pisces is easier to find in a dark sky, away from the light pollution of cities.

Deep-sky objects

Stargazers are spoiled for choice in this region of the sky. It has amazing large galaxies, such as NGC 488, and the NGC 520 system, which is a pair of colliding spiral galaxies that are slowly becoming one. These are best seen through a telescope. Another object it is worth trying to spot is M74 – a single beautiful spiral galaxy, similar to the Milky Way.

Brightest star

Eta Piscium, or Alpherg, was first discovered to be a binary star system in 1878. Its two stars have an orbital period of 850 years, meaning that since the system was discovered, it has only completed 17 per cent of its orbit.

Area and main stars

889 square degrees with 18 main stars

PISCIS AUSTRINUS

The Southern Fish

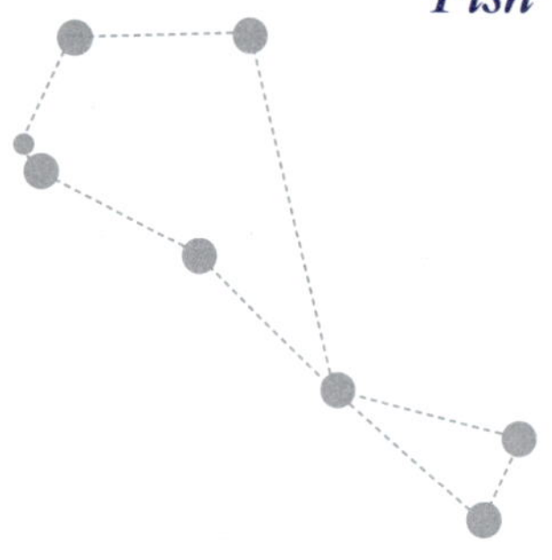

Lore

Piscis Austrinus is the southern hemisphere counterpart to Pisces. Both constellations represent a fish. This one originated from the Babylonian constellation simply known as the fish. It is thought that this constellation was introduced to the Ancient Greeks around 500 BCE. In Greek mythology, it is associated with the great fish that swallowed the water poured by Aquarius. In Egyptian mythology, the fish is said to have saved the life of the goddess Isis.

How to see it

Piscis Austrinus is a southern constellation, visible at latitudes between +55° and -90° and best viewed in October. It sits directly beneath the feet of Aquarius (p. 22). Find the head of the fish by looking for the constellation's brightest star, Fomalhaut.

Deep-sky objects

Using large telescope, or viewing from a small observatory, viewers can see the galaxies NGC 7172, which sits 110 million light-years from Earth, and NGC 7314, a large spiral galaxy.

Brightest star

Fomalhaut is one of the brightest stars in the Southern Sky. It is an A-type with a mass 1.9 times that of the Sun. It's estimated to be 440 million years old. It is suspected that Fomalhaut might be forming planets in a protoplanetary disk.

Area and main stars

245 square degrees with 7 main stars

SAGITTA

The Arrow

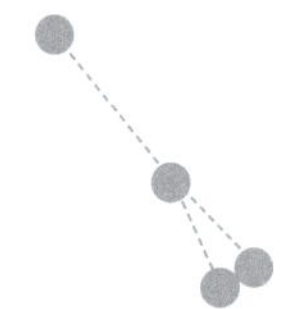

Lore

The story of the arrow in the sky is one of the oldest constellation myths. Its origins can be found in ancient Persian, Hebrew, Arabian, Armenian, Greek and Roman history. There are many different accounts about who fired this arrow, some attributing it to Cupid or Apollo. However, the most common story revolves around it being a poisoned arrow shot by Heracles.

How to see it

Sagitta is a northern constellation, visible at latitudes between +90° and -70° and best viewed in August. It appears as a Y-shape. Find it by locating its two red stars, Gamma Sagittae and Delta Sagittae, which form the length of the arrow.

Deep-sky objects

Two must-see objects in Sagitta are NGC 6886 and M71. NGC 6886 is a dim planetary nebula with a bright inner shell and a fainter red outer shell. M71 is a globular cluster of thousands of stars grouped together.

Brightest star

Gamma Sagittae is a red giant that is approximately 2.35 billion years old. It is 274 light-years from Earth and has a radius 55 times that of the Sun.

Area and main stars

80 square degrees with 4 main stars

SAGITTARIUS

The Archer

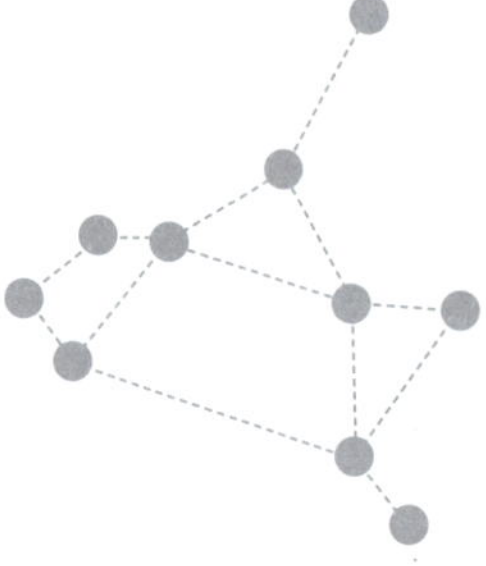

Lore

This constellation has represented many figures from different cultures and eras, including Nergal and Enkidu from Mesopotamia. The story that is often attributed to the modern representation of a centaur is the Greek myth of Chiron. Chiron was the child of the god Cronus and his niece Philyra. Chiron was half-man and half-horse and quickly became a favourite of the gods due to his gentle, wise temperament. Chiron was accidentally killed when Heracles mistakenly fired a poisoned arrow into him.

How to see it

Sagittarius is an equatorial constellation, visible at latitudes between +55° and -90° and best viewed in August. It is sometimes referred to as 'the tea pot'. Identify Scorpius (p. 100) and locate the bright red star Antares at its heart. Next, look left from the scorpion's tail and find the triangle of stars that form the spout of the tea pot.

Deep-sky objects

Sagittarius is home to many star clusters, such as M22 and M55, and nebulae, including the Lagoon and Trifid nebulae.

Brightest star

Epsilon Sagittarii is a binary star system. Its two stars – Epsilon Sagittarii A and B – are quite different from each other. Epsilon Sagittarii A is 3.80 times the mass of the Sun, and is spinning so rapidly – 243 kilometres per second – that it appears as an oval shape. Epsilon Sagittarii B's rotation is not well known, but the star is more similar to the Sun, with around 5 per cent less mass.

Area and main stars

867 square degrees with 10 main stars

SCORPIUS

The Scorpion

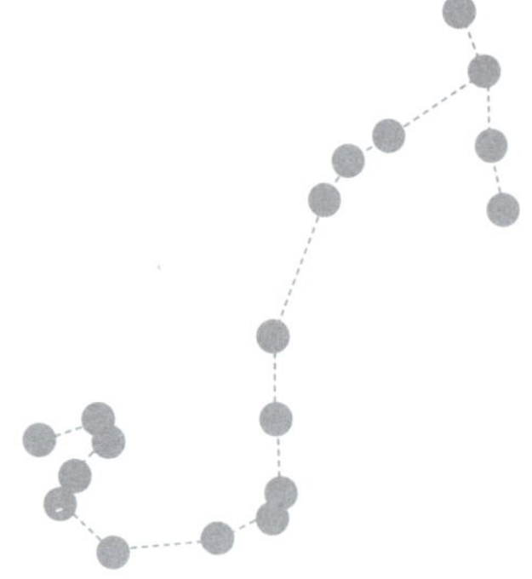

Lore

Scorpius has been recognised as the shape of a scorpion since 5000 BCE. In Greek mythology, one version of the story about Scorpius is that Hera was displeased with Orion, the hunter, and his hubris. She sent a giant scorpion to attack him. The scorpion was successful and after Orion's death, the gods placed it in the sky. The two combatants are now permanently separated by the cosmos.

How to see it

Scorpius is an equatorial constellation, visible at latitudes between +40° and -90° and best viewed in July. It is one of the easier constellations to spot, as it has many bright stars and a distinctive shape. Look for Antares, the very bright red star at its heart. From Antares, trace the tail of the scorpion's body to the south.

Deep-sky objects

The Butterfly Cluster NGC 6405 can be seen through binoculars and is roughly the same size in the sky as the full moon. As it name suggests, it is shaped like a butterfly with open wings.

Brightest star

Antares is often called the Heart of the Scorpion. It is the fifth-brightest star in the night sky and is actually a binary star system. The larger of the two stars is a red supergiant with a radius 680 times that of the Sun.

Area and main stars

497 square degrees with 18 main stars

SERPENS

The Snake

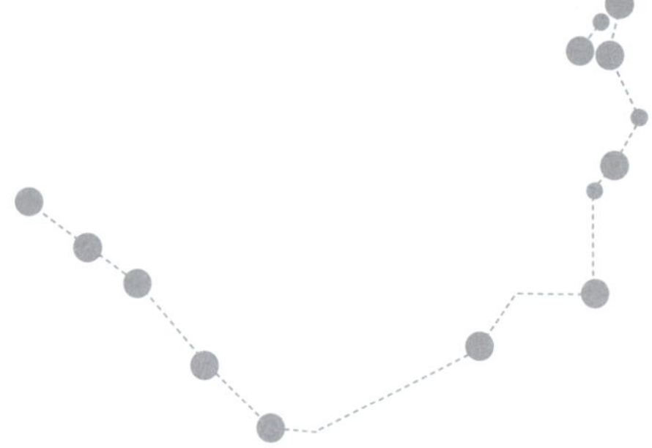

Lore

Serpens is the only constellation that is split into two parts: the head of the snake and the tail, which appear to be held by the constellation Ophiuchus. Greek mythology tells us that Asclepius (represented in the night sky by Ophiuchus) learned about life-saving herbs from a friendly snake who revived his friend. Ophiuchus and the snake were placed in the sky at the request of Apollo.

How to see it

Serpens is a northern constellation, visible at latitudes between +80° and -80° and best viewed in July. One section, Serpens Caput, is the head of the snake, and the other, Serpens Cauda, is the tail. First, locate Ophiuchus (p. 84). Serpens Caput appears to the left of Ophiuchus, and Serpens Cauda is to the right.

Deep-sky objects

This location is home to one of the sky's most famous objects: the Eagle Nebula, which contains the Pillars of Creation. The Pillars is a massive structure of gas and dust that is currently a stellar nursery, forming new stars. Viewers can also spot the globular cluster of stars NGC 5904.

Brightest star

Alpha Serpentis is a double-star system that is 74 light-years from Earth. It is over 2 billion years old and has evolved away from the main sequence. It is probably fusing helium in its core.

Area and main stars

637 square degrees with 11 main stars

TAURUS

The Bull

Lore

There are many myths about Taurus, but one that ties into many other constellations involves Orion, the hunter. In one story, it is said Orion wanted to prove he was the greatest hunter in the world by hunting all the wild animals. These actions angered the gods, and one god sent a giant scorpion to fight him. Not wanting the scorpion to be alone, they also sent the large bull Taurus to distract Orion. The ploy worked and Scorpius was the victor.

How to see it

Taurus is an equatorial constellation, visible at latitudes between +90° and -65° and best viewed in January. An easy way to identify it is to extend the line created by Orion's belt (p. 86) to the bright red star, Aldebaran, that forms one eye of the bull. The V-shape formed by the stars marks the face of the bull.

Deep-sky objects

Pleiades, also known as the Seven Sisters, has fascinated people worldwide for thousands of years. It is an open cluster that is fairly close to Earth. Taurus is also home to the spectacular Crab Nebula.

Brightest star

Alpha Tauri, or Aldebaran, is the fourteenth-brightest star in the night sky. It is a red giant with a radius 45 times that of the Sun. At 6.4 billion years old, this star is older than our solar system.

Area and main stars

797 square degrees with 11 main stars

TRIANGULUM

The Triangle

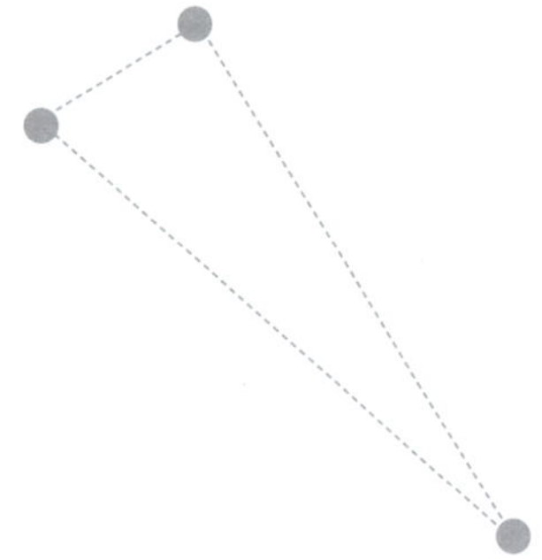

Lore

Triangulum is a small northern constellation. The stars within Triangulum were originally included in Babylonian star catalogues and formed part of a constellation known as the Plough, more commonly known now as the Big Dipper, which is located in Ursa Major. Triangulum is notable in history, as it was the first constellation listed on a pair of tablets compiling stars from 1000 BCE. The Ancient Greeks called Triangulum 'Deltoton' because it resembles the upper-case Greek letter delta (Δ). In Roman mythology, it represents the island Sicily. One Roman myth tells of Ceres, the patron goddess of Sicily, begging Jupiter to place the island in the heavens.

How to see it

Triangulum is a northern constellation, visible at latitudes between +90° and -60° and best viewed in December. It sits below the W-shape formed by Cassiopeia (p. 42) and west of Pleiades in Taurus (p. 104).

Deep-sky objects

One highlight in Triangulum is M33, known as the Triangulum Galaxy. It is 2.3 million light-years away and is a member of the local group we live in. It is just bright enough that, in dark skies, you can spot it with your naked eye. It will appear as a diffuse, fuzzy object. With a telescope, the spiral galaxies NGC 925 and NGC 672 are also visible.

Brightest star

Beta Trianguli is a binary star system located 127 light-years from Earth. The system is composed of two stars that are 2.6 and 2.25 times the mass of the Sun.

Area and main stars

132 square degrees with 3 mains stars.

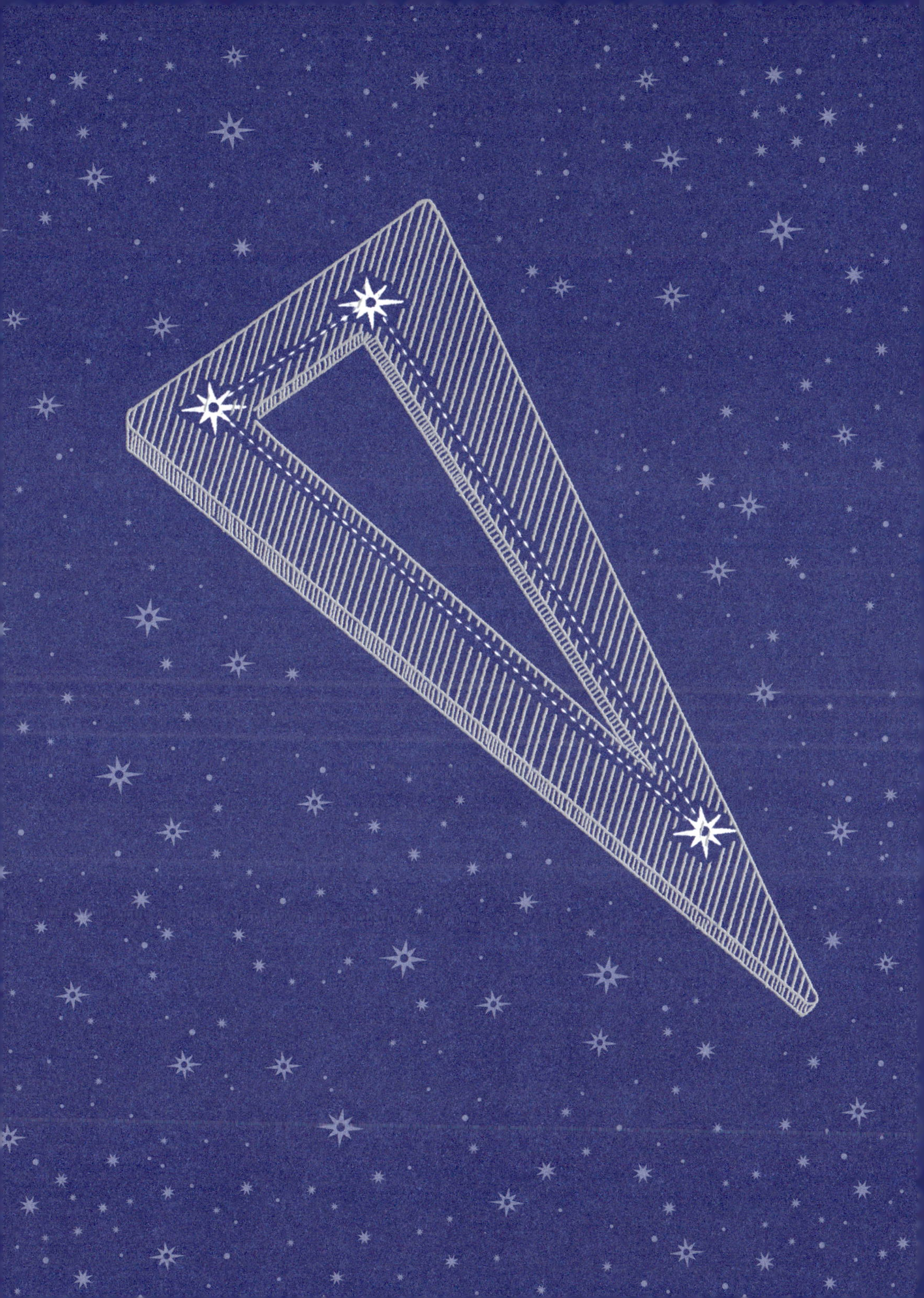

URSA MAJOR

The Great Bear

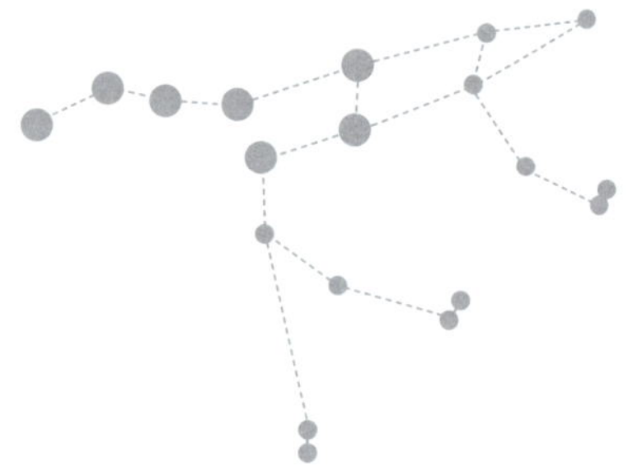

Lore

Ursa Major is associated with the myth of Hera, the wife of Zeus. Zeus cheated on Hera with Callisto, who bore him a son named Arcas. When Hera learned of Zeus's infidelity, she changed Callisto into the ugliest animal she could think of – a bear with shaggy fur, long teeth and claws. Callisto, unable to convince the townspeople who she really was, was banished to roam the forests.

How to see it

Ursa Major is a northern constellation, visible at latitudes between +90° and -30° and best viewed in April. It is home to the unofficial but popular asterism the Big Dipper, which makes it easy to find. The bear's tail contains three of the seven bright stars associated with the Big Dipper. The four stars that make up the Big Dipper's bowl form the backside and back leg of the bear.

Deep-sky objects

Several nearby galaxies can be seen in this region. One stargazing favourite is the Pinwheel Galaxy (M101). This galaxy is almost twice the radial size of our Milky Way and has more than 2 trillion stars.

Brightest star

Epsilon Ursae Majoris, or Alioth, appears along the tail of the bear. It is an A-type dwarf that is 2.9 times the mass of the Sun. For a long time, it was suspected to be two stars in a binary star system. However, further studies have not revealed a stellar companion, but do now suggest there might be a substellar object in orbit around Alioth.

Area and main stars

1280 square degrees with 7 main stars

URSA MINOR

The Little Bear

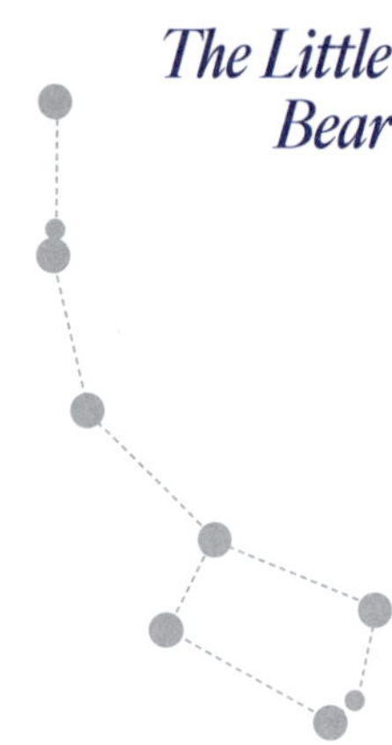

Lore

Ursa Minor, also known as the Little Dipper, has been associated with Arcas, the son of Zeus and Callisto. Myths tell that Hera turned Callisto into a great bear, doomed to roam the forests forever. Ursa Minor was once called Cynosura, after a nymph on Mount Ida in Crete who helped nurse the child Zeus while he was hidden from his father by his mother.

How to see it

Ursa Minor is a northern constellation, visible at latitudes between +90° and -10° and best viewed in June. It is one of the easiest constellations to find in the night sky. The tip of the bear's tail is formed by its brightest star, Polaris. The body is formed by four fainter stars that appear to rotate around Polaris.

Deep-sky objects

A few galaxies can be seen in the direction of Ursa Minor – NGC 6217, NGC 6251 and Ursa Minor Dwarf – but they are incredibly faint and require large telescopes.

Brightest star

Polaris is the brightest star in Ursa Minor. It is also well known as the North Star, because of its apparent location near the north celestial pole. Polaris is an evolved yellow supergiant with a mass 5.4 times that of the Sun. Polaris hasn't always been the North Star, due to Earth's precession. In 2750 BCE, the star Thuban, in the constellation Draco, was recorded as the north star.

Area and main stars

256 square degrees with 7 main stars

VIRGO

The Maiden

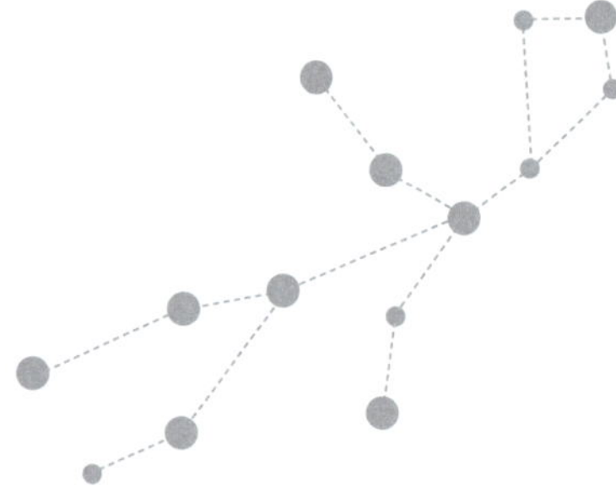

Lore

Virgo is often depicted as a woman holding a sheaf of wheat in her left hand. This refers to the Greek myth of Demeter, the goddess of agriculture. The legend tells of the kidnapping of Demeter's daughter, Persephone, by Hades, who wanted her to be his wife. Demeter was very unhappy about this, and refused to allow any crops to grow until a deal was struck. Persephone would spend six months of the year with her mother in Olympus and the remaining six months with Hades. The olive branch in Virgo's right hand comes from the myth of Astraea, and her promotion of peace and justice on Earth when other gods had fled back to the heavens.

How to see it

Virgo is an equatorial constellation, visible at latitudes between +80° and -80° and best viewed in May. First, locate Leo (p. 74). Virgo's head begins to form next to the lion's tail.

Deep-sky objects

Virgo is a fascinating region of the sky for stargazers, with more than thirty local galaxies visible. The Sombrero Galaxy (NGC 4594) is worth a look. Another galaxy easily seen through large binoculars is NGC 4579.

Brightest star

Spica appears as one star to the naked eye, but a telescope reveals that it is a binary star system. Its two stars orbit each other so closely that they look like one oval shape. They can only be separated by looking at the intensity of light they emit.

Area and main stars

1294 square degrees with 9 main stars

Modern constellations have been formed and crafted throughout history, but – unlike the ancients – they are often in parts of the sky with fainter stars. Some can be extremely hard to see from populated and light-polluted regions of the world.

Many of the modern constellations were mapped in the 15th and 16th centuries as Europeans began to explore the world by sea. Dutch–Flemish astronomer Petrus Plancius (1552–1622) identified twelve new constellations using the observations of the navigators on the first Dutch trade expedition to the East Indies in 1595. Pieter Dirkszoon Keyser (1540–1596) and Frederick de Houtman (1571–1627) were the first Europeans to see the stars in the uncharted skies they sailed under on that journey. They named many of the constellations they recorded after the strange new animals they saw on their travels.

French astronomer Nicolas-Louis de Lacaille (1713–1762) named all but one of his fourteen constellations for the scientific instruments developed during the Age of Enlightenment (1685–1815). The fourteenth constellation was Mensa, which he dedicated to Table Mountain in South Africa. He originally gave them French names, but they were renamed in Latin in the 1760s for consistency. Lacaille's work went far beyond charting the Southern Sky. He was also responsible for calculating the orbits of comets, and named Halley's Comet in honour of his fellow astronomer Edmond Halley (1656–1742).

Seven more modern constellations were catalogued by the Polish astronomer Johannes Hevelius (1611–1687). These were named for animals and objects.

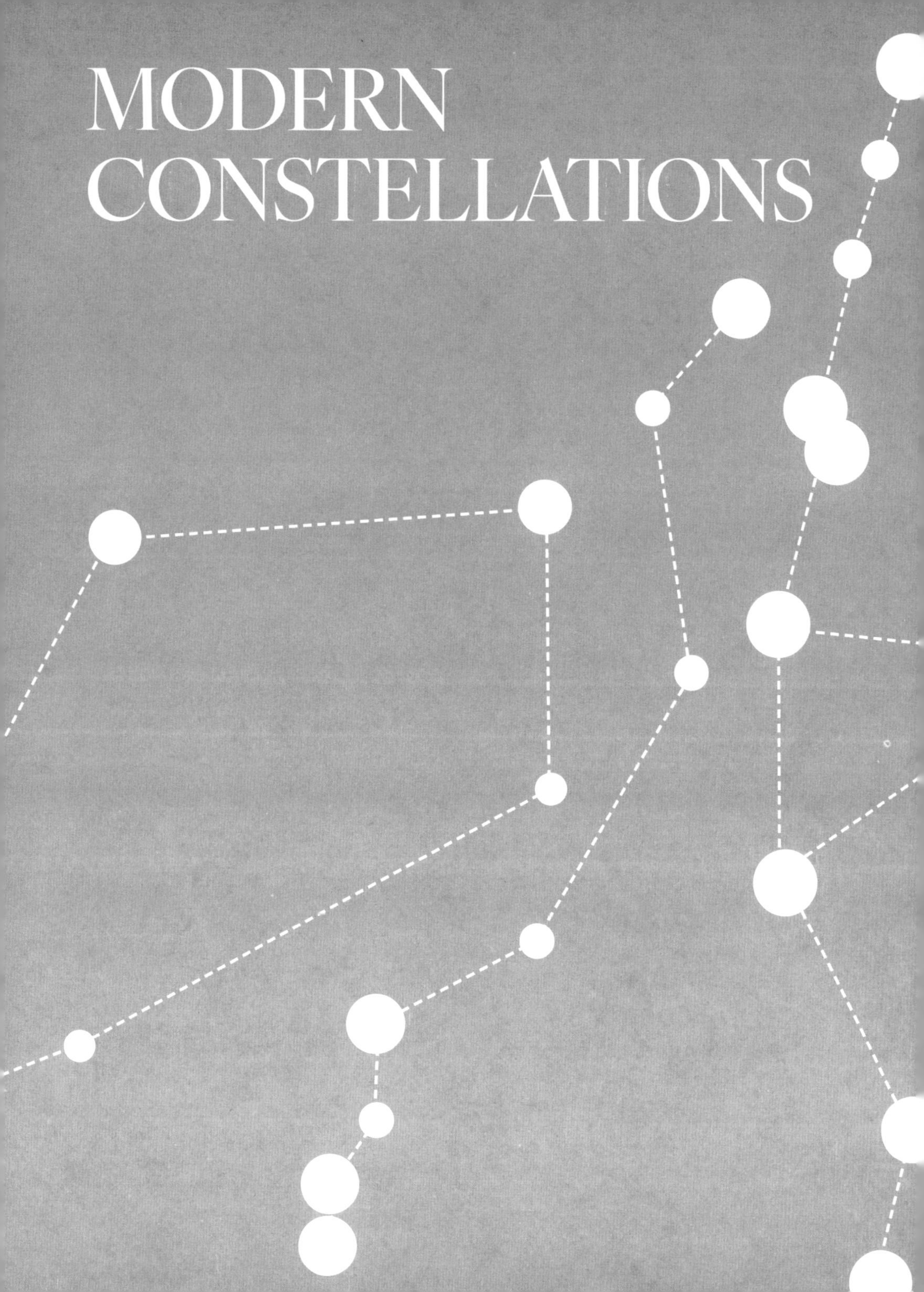
MODERN
CONSTELLATIONS

ANTLIA

The Air Pump

Lore

Nicolas-Louis de Lacaille (1713–1762) named this constellation Antlia Pneumatica for the air pump, an instrument invented by French physicist Denis Papin (1647–1712). Its name was shortened to Antlia in 1844.

How to see it

Antlia is a southern constellation, visible at latitudes between +45° and -90° and best seen in April. It is west of Centaurus (p. 44) and south of Hydra (p. 72). As it is a faint constellation, you may need to use a night-sky application to help you find it.

Deep-sky objects

Antlia contains many faint galaxies that can be seen through medium to large home telescopes. One of these is NGC 2997, a loosely wound, face-on spiral galaxy. This constellation also contains the Antlia Cluster, a large cluster of galaxies near our local group that is often studied by professional astronomers.

Brightest star

Alpha Antliae is a K-type giant located 320 light-years from Earth. It is best seen on a dark night, away from light pollution.

Area and main stars

239 square degrees with 3 main stars

APUS

The Bird of Paradise

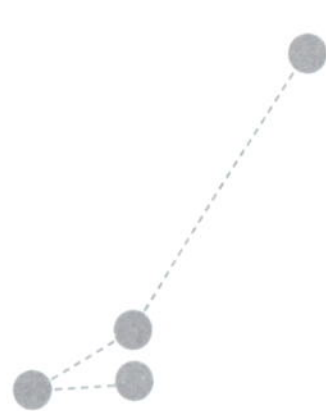

Lore

○ Apus is one of the modern constellations named by Pieter Dirkszoon Keyser (1540–1596). Apus represents the birds-of-paradise (*Paradisaea*), which is found across eastern Indonesia, Papua New Guinea and eastern Australia. At the time, it was believed that these birds had no feet ('apus' means 'footless' in Greek).

How to see it

✦ Apus is a southern constellation, visible at latitudes between +5° and -90° and best viewed in July. Locate the constellations Crux (p. 136) and Centaurus (p. 44). Draw a line from the bottom of Crux and the front feet of Centaurus. Apus is south of the point where the two lines meet.

Deep-sky objects

⊙ Apus contains two prominent globular clusters – NGC 6101 and IC 4499. NGC 6101 is 47,600 light-years from Earth and can only be seen using a large telescope. Research suggests that NGC 6101 contains many black holes.

Brightest star

⁂ Alpha Apodis has a radius 65 times that of the Sun and is 1000 times more luminous. It has consumed all the hydrogen in its core and has evolved to become a K-type giant. It is in the final stage of its life.

Area and main stars

◗ 206 square degrees with 4 main stars

CAELUM

The Engraver's Chisel

Lore

Originally called les Burins (the French expression for an engraver's chisel), like most of Lacaille's designations, this constellation was renamed in Latin in 1763. Les Burins became Caelum Sculptoris, which the English astronomer Francis Baily (1744–1817) later shortened to one word.

How to see it

✦ Caelum is a southern constellation visible at latitudes between +40° and -90° and best viewed in January. It is a smaller constellation with stars that appear faint to our eyes, so using a night-sky application will help. It lies between Eridanus (p. 66) and Dorado (p. 138).

Deep-sky objects

⊙ Due to its small size and location away from the galactic plane, there are few deep-sky objects visible near Caelum. There are no Messier objects, but there is one notable object – an unusual Seyfert galaxy called HE0459-2958. HE0459-2958 was first discovered as a jet: a large stream of material emanating from seemingly nothing. It is now thought that the jet is coming from a small galaxy.

Brightest star

⁂ Alpha Caeli is a double-star system made up of an F-type star 1.54 times the mass of the Sun, and an M-type star that is not visible to the naked eye.

Area and main stars

◗ 125 square degrees with 4 main stars

CAMELOPARDALIS *The Giraffe*

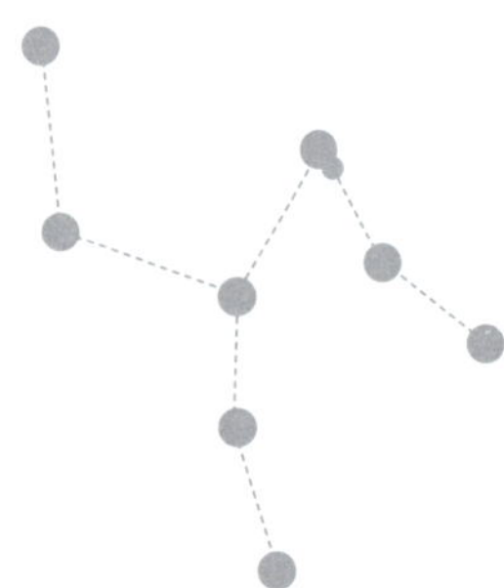

Lore

Camelopardalis is one of the newer constellations, first observed by Dutch astronomer Petrus Plancius (1552–1622) around 1612 and documented twelve years later by German astronomer Jakob Bartsch (1600–1633), who described it as a camel. The Greek word kamilopárdali (giraffe) translates to English as 'spotted camel' – a giraffe has a long neck like a camel and spots like a leopard.

How to see it

Camelopardalis is a northern constellation, visible at latitudes between +90° and -10° and best viewed in February. The stars in this region of the sky are so faint that the Ancient Greeks thought it was empty. Camelopardalis can be hard to find. Start at the Big Dipper – the seven bright stars in Ursa Major (p. 108). Trace the tip of the spoon shape near the head of the bear. On the other side of the sky, locate the W-shape in Cassiopeia (p. 42). Camelopardalis is directly between these two points.

Deep-sky objects

Camelopardalis is an excellent place to look for galaxies. Notable ones include NGC 2404, a spiral galaxy 12 million light-years away, and NGC 1569, an irregular dwarf galaxy.

Brightest star

Beta Camelopardalis is a G-type supergiant that is estimated to be 60 million years old. It is currently 6.5 times the mass of the Sun and has expanded in its old age to around 58 times the size of the Sun.

Area and main stars

757 square degrees with 8 main stars

CANES VENATICI *The Hunting Dogs*

Lore

The stars of Canes Venatici are not very bright and were not used in ancient constellations. They were listed by Greek astronomer Ptolemy as unfigured stars below the constellation Ursa Major. In 1533, German astronomer Petrus Apianus (1495–1552) depicted Boötes as having two dogs beside it. The dogs floated around in astronomical literature until 1687, when Polish astronomer Johannes Hevelius (1611–1687) decided to separate them from Boötes, identifying the constellation we now know as Canes Venatici.

How to see it

Canes Venatici is a northern constellation, visible at latitudes between +90° and -40° and best viewed in May. It lies between Ursa Major (p. 108) and Boötes (p. 32). As its stars are faint, it is best seen on a dark night, away from light-polluted areas.

Deep-sky objects

Canes Venatici contains five Messier objects, four of which are galaxies. One of the most notable is the Whirlpool Galaxy (M51), which is known for its interaction between a grand design spiral galaxy and a smaller galaxy. Other galaxies in Canes Venatici include NGC 4248, NGC 4242 and NGC 4707.

Brightest star

Cor Caroli, or Canum Venaticorum, is a binary star system. The first star is an A-type star 2.9 times the mass of the Sun that has been found to have a strong magnetic field. The second is an F-type star with a mass 1.47 times that of the Sun.

Area and main stars

465 square degrees with 2 main stars

CARINA

The Keel

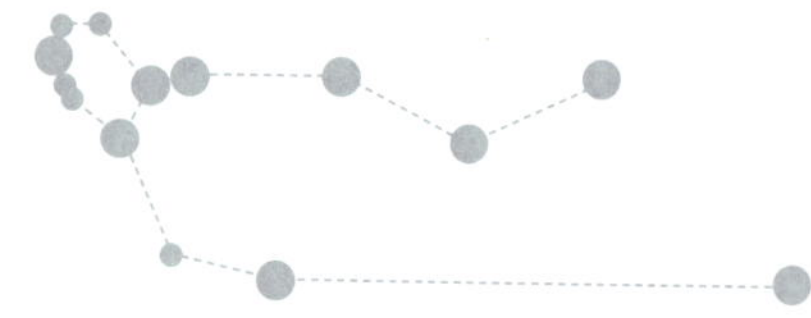

Lore

Carina was once a part of a larger constellation called Argo Navis, which, in Greek mythology, represented the ship the Argonauts used during their search for the Golden Fleece. The stars that made up Argo Navis were separated into three constellations – Carina, Puppis and Vela – each representing a different part of the ship. Carina is the keel.

How to see it

Carina is a southern constellation, visible at latitudes between +20° and -90° and best viewed in March. First, find Vela (p. 192) and Puppis (p. 174). Carina runs beneath these two constellations and covers their entire length. The brightest star marks the end of the ship's keel.

Deep-sky objects

This region of the sky has some of the most spectacular nebulae ever found, including the very popular Carina Nebula and Homunculus Nebula.

Brightest star

Canopus is the second-brightest star in the night sky. It is an A-type giant, located 310 light-years from Earth, with a luminosity 10,000 times that of the Sun. It is included in mythical lore across many cultures. To ancient Chinese astronomers, it was known as the 'Old Man of the South Pole'.

Area and main stars

494 square degrees with 9 main stars

CHAMAELEON

The Chameleon

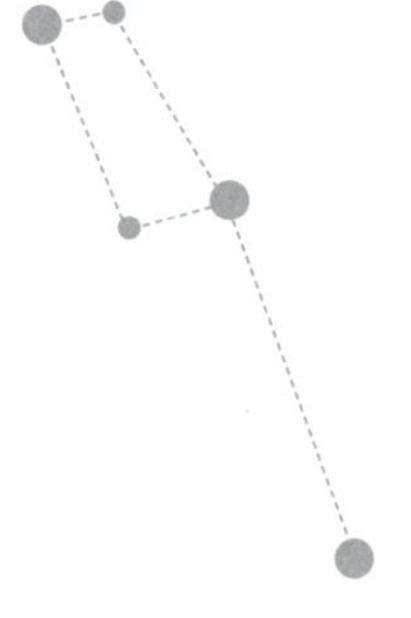

Lore

Chamaeleon is one of the twelve constellations described by Dutch navigators Pieter Dirkszoon Keyser (1540–1596) and Frederick de Houtman (1571–1627), who charted the Southern Sky on their expeditions. They named their constellations after animals they encountered on their travels. Like its namesake – the chameleon lizard – this constellation's faint stars make it blend in with its surroundings. It first appeared on a 1597–98 celestial globe.

How to see it

Chamaeleon is a southern constellation, visible at latitudes between +5° and -90° and best viewed in April. Look for the Large Magellanic Cloud, a satellite galaxy of the Milky Way that looks like a large dust cloud. Next, locate Crux (p. 136). Chamaeleon sits between them. A night-sky application will be useful.

Deep-sky objects

A recently discovered open cluster named Mamajek 1 sits in Chamaeleon. It is only 8 million years old and has twenty stellar members. Chamaeleon also contains a number of molecular clouds that are actively forming low-mass stars, and the planetary nebula NGC 3195.

Brightest star

Alpha Chamaeleontis is an F-type main sequence star. It is 1.8 billion years old and 1.4 times the mass of the Sun. It is currently 63.8 light-years from Earth but is steadily moving towards us. In 660,000+ years, it is expected to come within 47 light-years of our solar system.

Area and main stars

132 square degrees with 3 main stars

CIRCINUS

The Compass

Lore

Lacaille named this constellation le Compas and depicted it as a pair of diving compasses. It was renamed Circinus in 1763. In traditional Chinese astronomy, Circinus is in a region of the sky that symbolises the Azure Dragon of the East, which represents spring.

How to see it

Circinus is a southern constellation, visible at latitudes between +30° and -90° and best viewed in June. Although it is small, it is easy to find with the help of Crux (p. 136) and Centaurus (p. 44). Circinus sits south of Alpha Centauri and points towards the bottom star of Crux.

Deep-sky objects

Within Circinus are three open clusters and a planetary nebulae than can be seen with home telescopes. The clusters include NGC 5823, which is 800 million years old, NGC 5715, age unknown, and Pismis 20, a 5-million-year-old group. The planetary nebula NGC 5315 – first discovered in 1883 – is also visible using larger telescopes.

Brightest star

Alpha Circini is a variable A-type star. Its brightness changes fairly quickly but it is not noticeable to the naked eye. This star is 1.5 times the mass of the Sun and 10 times more luminous.

Area and main stars

93 square degrees with 3 main stars

COLUMBA

The Dove

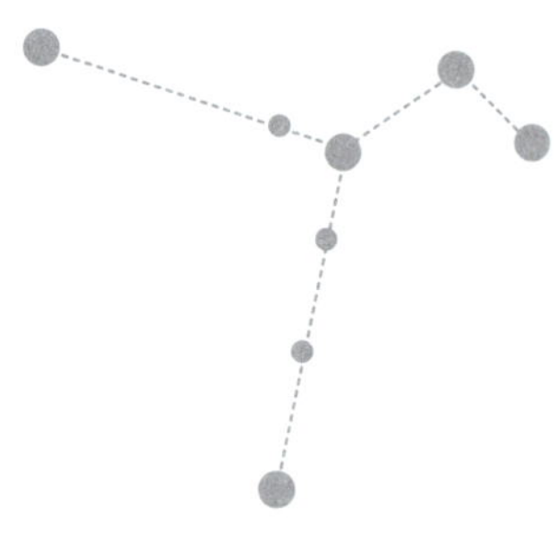

Lore

This constellation was originally named Columba Noachi by Plancius. The name referred to the biblical story of a dove bringing an olive branch to Noah to tell him the great flood was receding. The name was shortened to Columba when German cartographer Johann Bayer (1572–1625) included it in his 1603 star atlas. Although Columba is recognised as a fairly new constellation, the stars within it are described in Aratus's poem *Phainomena*, which was written in the 3rd century BCE.

How to see it

Columba is a southern constellation, visible at latitudes between +45° and -90° and best viewed in January. Locate Sirius, the brightest star in the sky and part of Canis Major (p. 36), and Canopus, the brightest star in Carina (p. 126). Columba sits between these two.

Deep-sky objects

There are multiple deep-sky objects within Columba, including galaxies and open globular clusters. The brightest object is NGC 1851, a massive globular cluster. It sits 39,000 light-years from Earth and is estimated to be 9.2 billion years old and have more than 551,000 times the mass of the Sun. Other objects include galaxies NGC 1808, NGC 1792 and NGC 2090.

Brightest star

Alpha Columbae is 261 light-years from Earth. It is a B-type star surrounded by a hot gas disk. Its traditional name, Phact, comes from the Arabic word for 'ring dove'.

Area and main stars

270 square degrees with 5 main stars

COMA BERENICES

Berenices Hair

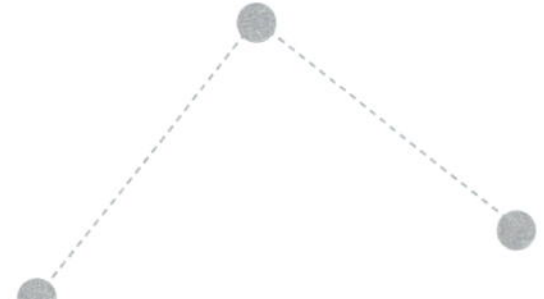

Lore

Coma Berenices is the first of the modern constellations, and the only one to be named for a historical figure. Queen Berenice II of Egypt sacrificed her hair to the goddess Aphrodite to ensure her husband's safe return from battle.

How to see it

Coma Berenices is a northern constellation, visible at latitudes between +90° and -70° and best viewed in May. Find Leo (p. 74) and draw a line from the star Zosma to the red star Arcturus in Boötes (p. 32). Coma Berenices is in the middle of this line. The top centre of this constellation forms an upside-down 'Y'.

Deep-sky objects

There are many galaxies in Coma Berenices that are visible through binoculars or small telescopes, including the spiral galaxies NGC 4450 and M100. Also in this location is the Coma Cluster, a cluster of galaxies that sit near each other, some even interacting.

Brightest star

Despite being a G-type dwarf, Beta Comae Berenices is the brightest star in this constellation. Its mass is just 1.15 times greater than the Sun, which is also a G-type dwarf. Although we haven't detected any planets around this star yet, if an Earth-like planet were to be detected, its habitable zone would be similar to our own.

Area and main stars

386 square degrees with 3 main stars

CRUX

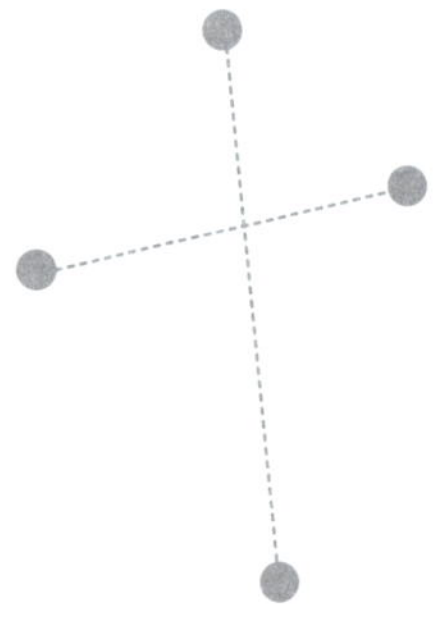

Lore

Crux was originally seen as part of Centaurus by the Ancient Greeks, as it sits low to the horizon and was visible from the Northern Hemisphere for parts of the year. Over time, the gradual shift in Earth's rotation meant that Crux fell below the horizon and could no longer be seen by most Europeans. It was 'rediscovered' by European navigators in the 16th century. In some Indigenous language groups of Australia, Crux has its own mythology. The Boorong people tells of Bunya, the hunter, who was chased by Tchingal, the celestial emu.

How to see it

Crux is a southern constellation, visible at latitudes between +20° and -90° and best viewed in May. Although small, it is very bright, making it one of the easiest constellations to find. It can often be found simply by looking for the iconic four stars that form the shape of a cross. A fifth, fainter star sits to the right. Crux can also be found by looking for the two bright stars that form part of Centaurus (p. 44), which point directly to it.

Deep-sky objects

Crux contains two interesting objects: the Coalsack Nebula (a dark nebula) and the Jewel Box (NGC 4755).

Brightest star

Alpha Cru, or Acrux, is the star at the bottom of the cross. To the naked eye, it looks like one star, but with a telescope it appears as a triple star system. It is actually a six star system, with four B-type massive blue stars.

Area and main stars

68 square degrees with 4 main stars

DORADO

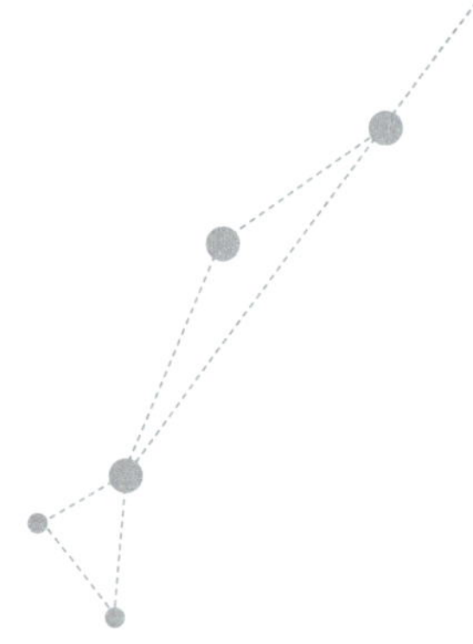

Lore

Dorado is one of the newer constellations. It was first observed and mapped by Plancius. The name means 'dolphinfish' in Spanish, but it has also been referred to as a swordfish.

How to see it

Dorado is a southern constellation, visible at latitudes between +20° and -90° and best viewed in January. It is one of the smaller constellations but can be found by locating the Large Magellanic Cloud. To the north are three stars that appear to be in a line. These are the stars that form this constellation.

Deep-sky objects

Stargazers can take a close look at the Large Magellanic Cloud with binoculars or small telescopes. Within it is the Tarantula Nebula, in which a supernova was witnessed in 1987 – the closest one we've seen in modern times.

Brightest star

Alpha Doradus is one of the brightest binary star systems in the night sky. It consists of a B-type subgiant orbiting around an A-type giant.

Area and main stars

179 square degrees with 3 main stars

FORNAX

The Furnace

Lore

Fornax was originally named le Fourneau Chimique by Lacaille, for the chemical furnace. It became Fornax Chimiae in 1763, which was later shortened to Fornax.

How to see it

Fornax is a southern constellation, visible at latitudes between +50° and -90° and best viewed in December. It is a smaller constellation that fills the space between parts of Eridanus (p. 66). Fornax sits north of Achernar, which is one of the brightest stars in the Southern Sky. A night-sky application will help you find it. It is best seen away from light-polluted areas.

Deep-sky objects

Members of our local group of galaxies can be seen in Fornax, including the Fornax Dwarf galaxy. This small galaxy sits 460,000 light-years from Earth and is a satellite of the Milky Way. It can be viewed through larger amateur telescopes. Also visible is NGC 1097, a barred spiral galaxy that is 45 million light-years from Earth.

Brightest star

Alpha Fornacis is a triple star system. The first star, Fornacis A, is an F-type subgiant that is 1.3 times the mass of the Sun and 2.9 billion years old. Fornacis B is a blue straggler, which is an evolved star that appears bluer and brighter than it should, probably because it is consuming material from a nearby star. Fornacis B is orbited by Fornacis Bb, which is probably a white dwarf.

Area and main stars

398 square degrees with 2 main stars

GRUS

The Crane

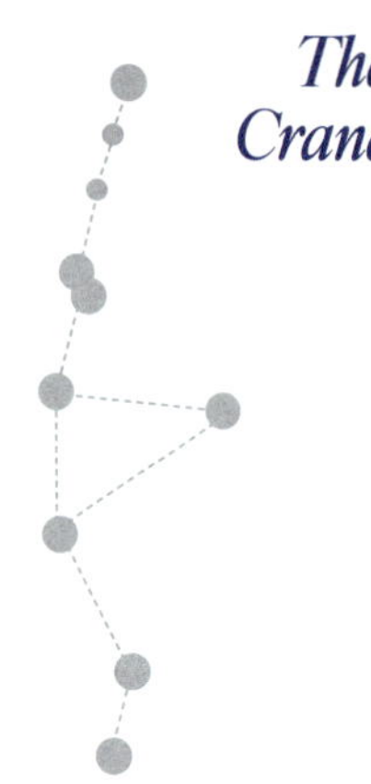

Lore

Grus was originally part of the neighbouring constellation Piscis Austrinus and was seen as part of the fish's tail. Grus was defined as a separate constellation by Plancius, and it first appeared on the 1598 celestial sphere. It was also recorded on the 1603 celestial atlas created by Johann Bayer. In the 17th century, it was briefly renamed Phoenicopterus, which is Latin for 'flamingo'. However, this name didn't stick and it is still known as Grus today.

How to see it

Grus is a southern constellation, visible at latitudes between +34° and -90° and best viewed in October. It sits between Piscis Austrinus (p. 94) and Tucana (p. 190). Look for the triangle made by its three brightest stars, two of which appear slightly red.

Deep-sky objects

There are multiple deep-sky objects in Grus, including IC 5148, a planetary nebula that sits 3000 light-years from Earth and has an expanding outer shell. Recent studies show that material is moving outwards from its former star at 50 kilometres per second. Other objects include the Grus Quartet – four galaxies (NGC 7552, NGC 7590, NGC 7599 and NGC 7582) near each other.

Brightest star

Alpha Gruis, or Alnair, is one of the brightest stars in the night sky. It is one of the fifty-eight stars used for celestial navigation. It is a single B-type star that is 3.8 times the mass of the Sun. It is a young star – just 100 million years old. If a Stegosaurus ever looked at the night sky, Alnair would not have been there.

Area and main stars

366 square degrees with 8 main stars

HOROLOGIUM

The Pendulum Clock

Lore

In 1750, Lacaille led an expedition to the Cape of Good Hope, where he spent two years cataloguing almost 10,000 stars in the Southern Sky. During this time, he mapped many new constellations and named fourteen of them. He originally described Horologium as a clock with a pendulum and a second hand (l'Horloge à pendule & à secondes). The name became Horologium in 1763.

How to see it

Horologium is a southern constellation, visible at latitudes between +30° and -90° and best viewed in December. Locate the Large Magellanic Cloud, the Small Magellanic Cloud and Orion (p. 86). Horologium sits between these three. It appears about half the size of Orion, but due to its faint stars, it is not as easily identified.

Deep-sky objects

Many deep-sky objects can be found with telescopes, including the globular clusters NGC 1261 and Arp-Madore 1. At 402,000 light-years away, Arp-Madore 1 is the most distant known globular cluster in the Milky Way's halo. Stargazers may also spot NGC 1512, a barred spiral galaxy located 38 million light-years from the Milky Way.

Brightest star

Alpha Horologii is a K-type red giant. It is 1.4 times of the mass of the Sun, and 3.5 billion years old. At some point it migrated away from the main sequence, and it has expanded its outer layers to almost 10 times that of the Sun.

Area and main stars

249 square degrees with 6 main stars

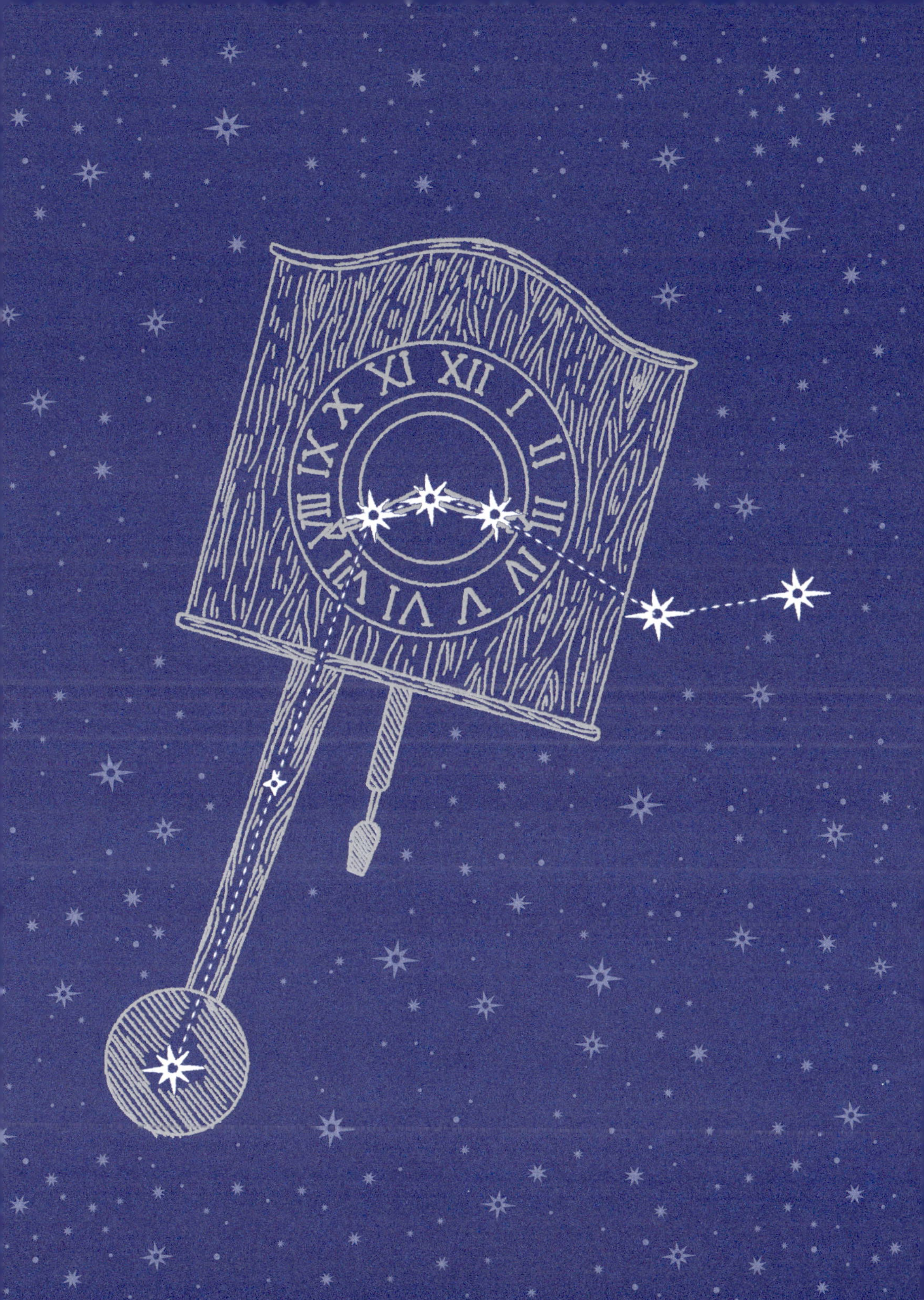

HYDRUS

The Water Snake

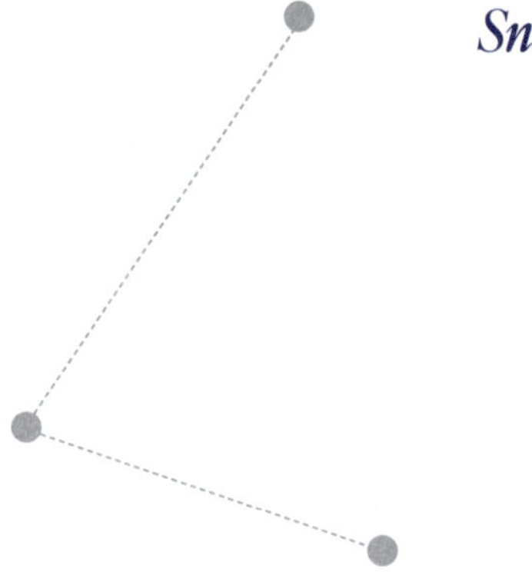

Lore

Hydrus is one of the twelve constellations first described by Plancius. Hydrus was originally called De Waterslang for the water snakes that Keyser and Houtman encountered on the first Dutch trade expedition to the East Indies. The name was changed to l'Hydre Mâle, to distinguish it from the constellation Hydra, before it was Latinised to Hydrus.

How to see it

Hydrus is a southern constellation, visible at latitudes between +8° and -90° and best viewed in December. It passes between the Small and Large Magellanic Clouds. Locate the very bright star Achernar in Eridanus (p. 66). Next, look south to find the next brightest star. This star marks the head of the snake.

Deep-sky objects

Only faint deep-sky objects can be seen in Hydrus, and most galaxies will require medium to large telescopes. Galaxies include the irregular galaxy NGC 1473 and the spiral galaxy NGC 1511.

Brightest star

Beta Hydri is a main sequence star, approximately 1.12 times the mass of the Sun and slightly older, at an estimated age of 6+ billion years.

Area and main stars

243 square degrees with 3 main stars

INDUS

The Indian

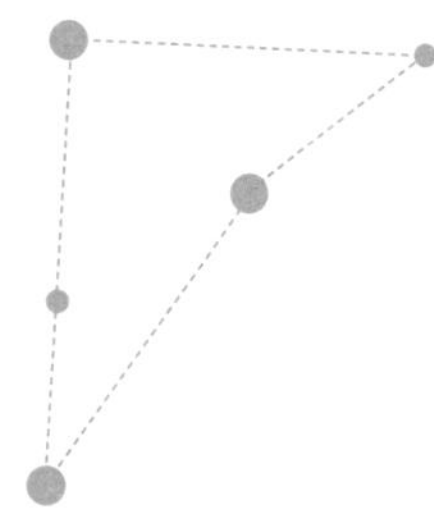

Lore

Indus was recorded by Plancius and the name is thought to refer either to Native Americans or to Indigenous people from the parts of Asia being explored by Europeans at the time. The constellation itself is relatively faint. It has typically been depicted as a man holding arrows or spears.

How to see it

Indus is a southern constellation, visible at latitudes between +15° and -90° and best viewed in September. It lies west of the Small Magellanic Cloud. Using a night-sky application in an area far from light pollution will help stargazers find and identify the stars of this constellation.

Deep-sky objects

There are multiple nearby galaxies, including NGC 7038, a large spiral galaxy located 210 million light-years away; NGC 7049, a lenticular galaxy 100 million light-years away; and NGC 7090, an edge-on spiral galaxy only 31 million light-years away.

Brightest star

Alpha Indi is a K-type star, halfway between a subgiant and a giant. It is 2.15 times the mass of the Sun and might have two companion M-type dwarf stars, but that hasn't been confirmed as of 2025.

Area and main stars

294 square degrees with 3 main stars

LACERTA

The Lizard

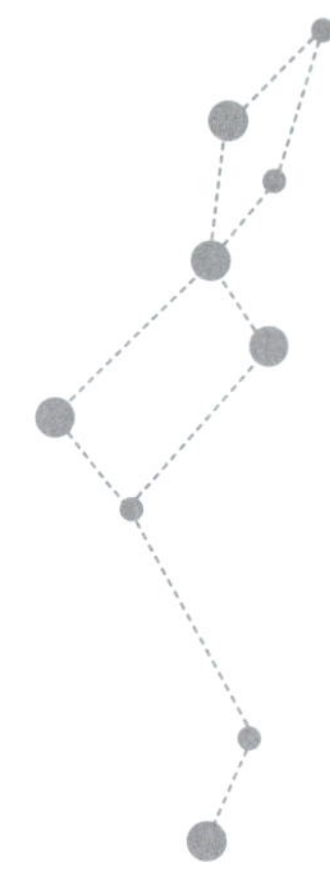

Lore

No classic mythology is associated with Lacerta, as it is one of the constellations named by Hevelius. It was initially named Stellio after a lizard found along the Mediterranean coast. In Chinese astronomy, the northern part of Lacerta was part of the figure Tengshe, a flying serpent.

How to see it

Lacerta is a northern constellation, visible at latitudes between +90° and -40° and best viewed in October. Lacerta can be tricky to see because it is primarily made up of very faint stars. Locate Andromeda (p. 20), then use a night-sky application to navigate from there. Lacerta might not be visible in areas with light pollution.

Deep-sky objects

The open cluster NGC 7243 can be seen with a telescope. This cluster is just over 1 billion years old. Another object, often mistaken for a star, is BL Lacertae, which is the core of a distant galaxy.

Brightest star

Alpha Lacertae is 103 light-years from Earth. It is actively burning hydrogen in its core, and is 400 million years old. Alpha Lacertae rotates relatively quickly – 67 times the rate of the Sun.

Area and main stars

201 square degrees with 5 main stars

LEO MINOR

The Smaller Lion

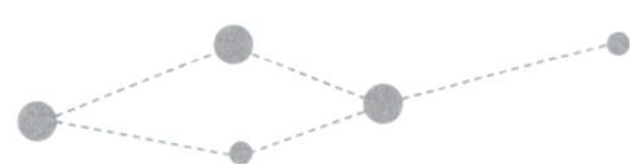

Lore

To the Greek astronomers Aratus and Ptolemy, this area of the sky did not contain any distinctive patterns. It wasn't until 1687 that Hevelius first depicted Leo Minor as aligning with neighbouring constellations Leo and Ursa Major. German astronomer Christian Ludwig Ideler (1766–1846) theorised that the stars of Leo Minor could be seen on a 13th-century Arabic celestial globe, on which it was referred to as 'Gazelle with her Young'.

How to see it

Leo Minor is a northern constellation, visible at latitudes between +90° and -45° and best viewed in April. It lies between Leo (p. 74) and Ursa Major (p. 108). Find the bright stars Regulus in Leo and Dubhe in Ursa Major and draw a line between them. The tail of Leo Minor sits in the middle of the line.

Deep-sky objects

Leo Minor contains many deep-sky objects, but not all are visible through home telescopes. NGC 3414 is a lenticular galaxy 77 million light-years from Earth, and NGC 3504 is a spiral galaxy 88 million light-years away. The Leonis Minorid meteor shower, which appears to come from Leo Minor, occurs between 18 October and 29 October and is visible in the Northern Hemisphere.

Brightest star

46 Leonis Minoris, also called Praecipua, is a K-type star roughly the mass of the Sun but with a radius 8.2 times larger. It is red clump star, meaning it is slightly hotter than most red giants.

Area and main stars

232 square degrees with 3 main stars

LYNX

The Lynx

Lore

This constellation was mapped in 1687 by Hevelius to define the sky between Ursa Major and Auriga. He named it Lynx because it is so faint that it is very hard to see. He reportedly challenged future stargazers to spot it, declaring that they would need to be 'lynx-eyed' to find it. It was also referred to as Tigris, but the official name was recorded as Lynx in the catalogues.

How to see it

Lynx is a northern constellation, visible at latitudes between +90° and -55° and best viewed in February. It is very faint and can be challenging to see without a night-sky application.

Deep-sky objects

In Lynx, you can spot the most distant globular cluster within our galaxy, NGC 2419, also called the Intergalactic Wanderer. It is estimated to be 275,000+ light-years from Earth. It is theorised that this cluster is in a highly elliptical orbit around the Milky Way. Viewers can also spot NGC 2537, a blue compact dwarf galaxy somewhere between 17 and 30 million light-years away.

Brightest star

Alpha Lyncis is a K-type red giant with a mass 1.5 times that of the Sun and a radius 58 times larger.

Area and main stars

545 square degrees with 4 main stars

MENSA

The Table Mountain

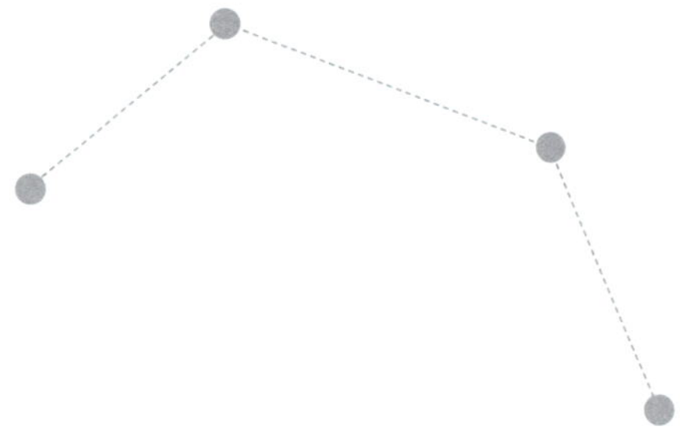

Lore

Mensa is the only one of Lacaille's fourteen constellations that he didn't name after a scientific instrument. He originally named this region of the sky Montagne de la Table, in honour of South Africa's Table Mountain. The English astronomer John Herschel (1792–1871) proposed shrinking the name to Mensa, which is Latin for 'table'. Mensa borders on the bottom of the Large Magellanic Cloud, a dwarf galaxy of the Milky Way. The Large Magellanic Cloud is known as Brolga by the Kamilaroi and Euahlayi people of Australia.

How to see it

Mensa is a southern constellation, visible at latitudes between +4° and -90° and best seen in January. Although not large, it is one of the easiest to locate in the sky. It is slightly south of the Large Magellanic Cloud.

Deep-sky objects

The Large Magellanic Cloud is the most well-known deep-sky object in this region of the sky, and it is visible to the naked eye. Features within the cloud can be explored using binoculars or small telescopes. Also within Mensa are NGC 1987, a 600-million-year-old globular cluster, and NGC 1848, a 27-million-year-old open cluster.

Brightest star

Alpha Mensae is a faint star that is best seen in dark skies away from light pollution. It is a G-type star with a slightly smaller mass and radius than the Sun.

Area and main stars

153 square degrees with 4 main stars

MICROSCOPIUM

The Microscope

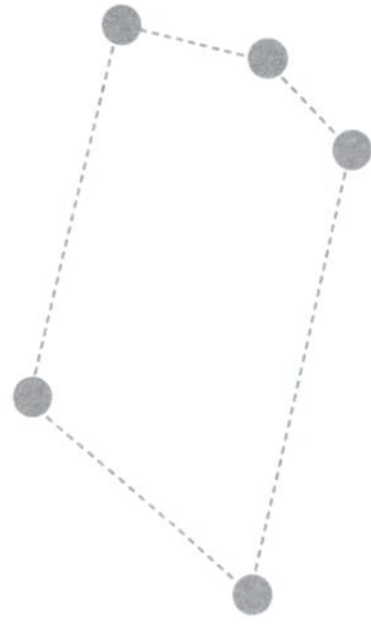

Lore

Microscopium was first described and named by Lacaille. The invention of the microscope was the culmination of over 4000 years of investigation involving lenses and water-filled spheres. The name 'microscope' was coined in 1625, after Galileo Galilei (1564–642) unveiled his version of the compound microscope.

How to see it

Microscopium is a southern constellation, visible at latitudes between +45° and -90° and best viewed in September. As it is a very faint constellation, a night-sky application used far away from light-polluted areas will be helpful.

Deep-sky objects

English astronomer Patrick Moore (1923–2012) described Microscopium as 'totally unremarkable', and said it held nothing of interest. But every square inch of the night sky has beauty and wonder hidden within it. NGC 6925 – a spiral galaxy 99 million light-years away – is visible through large home telescopes. The Microscopids minor meteor shower also radiates from this region in June/July.

Brightest star

Gamma Microscopii is too faint to see in light-polluted skies. It is a G-type giant with a mass 2.5 times that of the Sun, located 223 light-years from Earth. Although it is a faint star today, it has been suggested that it is part of a larger group of stars that have moved through the galaxy in a traceable motion. If that is true, 3.9 million years ago, it would have only been 6 light-years from Earth, making it brighter in the night sky than Sirius is now.

Area and main stars

210 square degrees with 5 main stars

MONOCEROS

The Unicorn

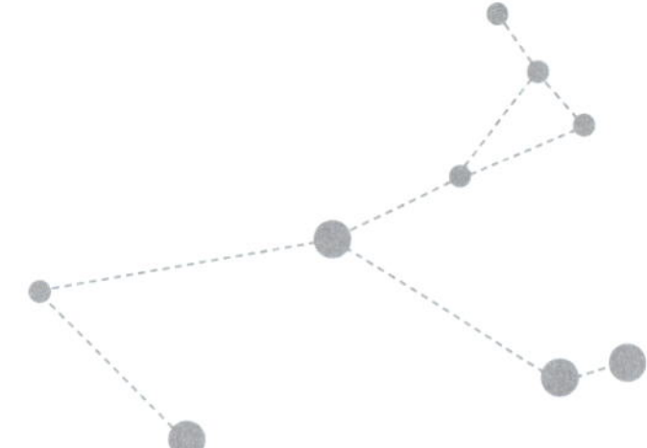

Lore

Monoceros is a fairly new constellation, first appearing in 1612, on a globe created by Petrus Plancius. The imagery of a unicorn might refer to a creature mentioned in the Bible, which is often seen as an emblem of purity.

How to see it

Monoceros is a northern constellation, visible at latitudes between +75° and -90° and best viewed in February. Although it is a faint constellation, it can be easily located using Orion (p. 86). Find Betelgeuse in Orion's right shoulder, then look to the left to see the stars that form the unicorn's head. The legs of the unicorn end near the legs of Orion.

Deep-sky objects

The open cluster M50 sits between the front and back legs of Monoceros and is visible through binoculars. Another object that is visible through home telescopes is V616 Monocerotis. This seemingly unremarkable star is in a binary star system with a stellar-mass black hole – one of the closest to Earth! Although the black hole isn't visible, it is incredible to think about it as you observe its star companion.

Brightest star

Alpha Monocerotis is an evolved giant. It is classified as a red clump star, and is generating energy through helium fusion at its core. It is 890 million years old.

Area and main stars

482 square degrees with 4 main stars

MUSCA

The Fly

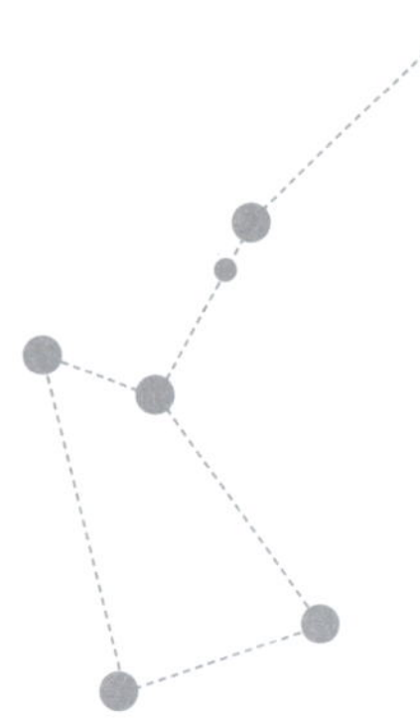

Lore

Musca, the fly, is a small constellation in the Southern Sky that was first identified in 1598 by Plancius. Musca was originally known as De Vlieghe (Dutch for 'the fly') and was later referred to in Johann Bayer's catalogue as Apis, or 'the bee'. In Australia, the Wardaman people perceive the main stars of Musca as a ceremonial boomerang in a sacred region around Crux.

How to see it

Musca is a southern constellation, visible at latitudes between +10° and -90° and best viewed in May. First find Crux (p. 136). Musca sits directly beneath the bottom of the cross, and its brightest stars form an irregular quadrilateral.

Deep-sky objects

The planetary nebula MyCn 18, also known as the Engraved Hourglass Nebula, is visible through larger home telescopes. It was discovered by Annie Jump Cannon (1863–1941) and Margaret Mayall (1902–1995) in the early 20th century. The centre looks like an eye, and rings of gas form the hourglass. Another object that can be seen with the naked eye is the dark spot of the Coalsack Nebula. In Indigenous Australian astronomy, it forms the head of the celestial emu.

Brightest star

Alpha Muscae is a young B-type subgiant. It has a mass 8.8 times greater than the Sun and is only 18 million years old. As a bigger star, it will burn through its hydrogen fuel quicker than stars like the Sun. It will probably die in a supernova explosion.

Area and main stars

138 square degrees with 6 main stars

NORMA

The Ruler

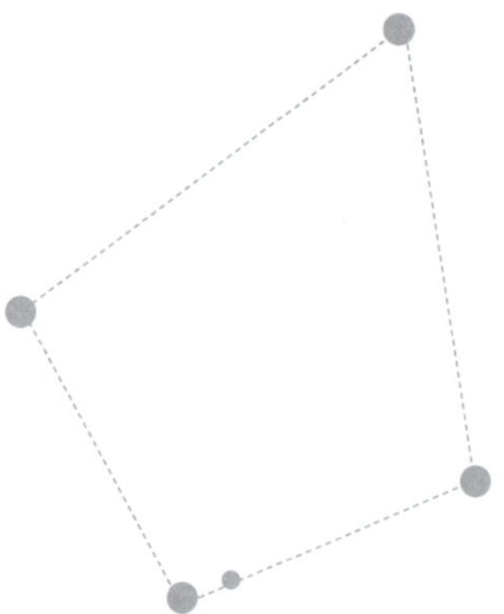

Lore

In 1752, Lacaille described this constellation and named it l'Équerre et la Règle ('the square and the rule'). This name was later shortened and Latinised to Norma.

How to see it

Norma is a southern constellation, visible at latitudes between +30° and -90° and best viewed in June. It is a small constellation that sits between two giants, Centaurus (p. 44) and Scorpius (p. 100). Find Alpha Centauri, the brightest star in Centaurus, and the red-hued star Antares in Scorpius. These two stars often stand out in the night sky. Norma sits about halfway between them, slightly closer to the galactic plane.

Deep-sky objects

As Norma is close to the galactic plane and centre, there are many deep-sky objects to be seen, including star clusters. NGC 6087 is the brightest cluster visible in this constellation. It is 100 million years old and located 3300 light-years from Earth. Another object is Mz 3, known as the Ant Nebula, which is 3500 light-years away. This is a bipolar planetary nebula – one of the rarest types of nebulae, recognisable by their distinct butterfly or hourglass shape. It is a spectacular sight.

Brightest star

Gammae-2 Normae is a K-type yellow giant. It is 120 light-years from Earth but is moving towards our solar system at 29 kilometres per second. Over the next 10 or 20 million years, it will start to appear brighter in the night sky.

Area and main stars

165 square degrees with 4 main stars

OCTANS

The Octant

Lore

Octans is the constellation that covers the south celestial pole. Lacaille originally named it l'Octans de Reflexion (French for 'the reflecting octant'). An octant is a navigational instrument that was invented in the 1730s. The constellation's name was eventually shortened to Octans.

How to see it

Octans is a southern constellation, visible at latitudes between +0° and -90° and best viewed in October. One way to find it is to locate the Large and Small Magellanic Clouds, and Alpha Centauri in Centaurus (p. 44). Octans sits between the three. Another fun thing to try is to stargaze for multiple hours and watch as the stars appear to rotate around the central point of Octans.

Deep-sky objects

Multiple galaxies are visible through medium to large home telescopes. NGC 2573 is a faint spiral galaxy that is close to the south celestial pole. You may also spot NGC 7095 and NGC 7098, spiral galaxies that are 115 million and 95 million light-years away, respectively.

Brightest star

Sigma Octantis is a spectroscopic binary star system. The main star is a K-type giant with a radius 5.8 times that of the Sun. The second is probably a red or white dwarf.

Area and main stars

291 square degrees with 3 main stars

PAVO

The Peacock

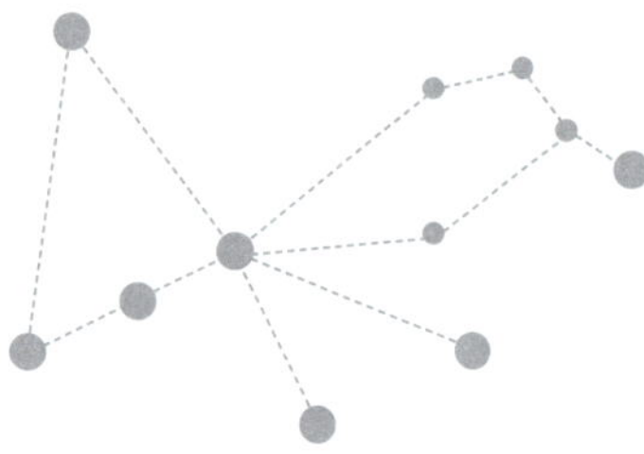

Lore

Keyser first identified Pavo in the late 16th century. Pavo represents a peacock – it may have been named for the unique birds that Keyser encountered on his first voyage to the East Indies.

How to see it

Pavo is a southern constellation, visible at latitudes between +30° and -90° and best viewed in August. It is easiest to spot in dark skies, away from the bright lights of big cities. Find the Small Magellanic Cloud, and Antares in Scorpius (p. 100). Draw a line between the two. Pavo sits one-third of the way along the line from the Small Magellanic Cloud. Its brightest star forms the head of the bird.

Deep-sky objects

Stargazers can spot stunning galaxies in Pavo, such as NGC 6744, a large spiral galaxy similar to the Milky Way, and NGC 6782 and NGC 6872, barred spiral galaxies with interesting shapes.

Brightest star

Alpha Pavonis is a binary star system. The main star is a blue subgiant that is 6 times the mass of the Sun. The properties of the second star are not well known.

Area and main stars

378 square degrees with 7 main stars

PHOENIX

The Phoenix

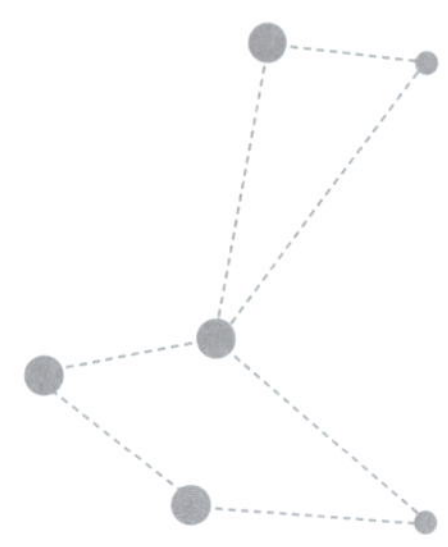

Lore

Phoenix is the largest of the twelve constellation that Petrus introduced. Unusually for a modern constellation, it was referenced in ancient mythology by Arabic astronomers, for whom it represented young ostriches, a griffin or an eagle.

How to see it

Phoenix is a southern constellation, visible at latitudes between +32° and -90° and best viewed in November. Find the Small Magellanic Cloud. Next, locate the bright star nearby known as Ankaa. This is the brightest star in Phoenix and forms the bird's head.

Deep-sky objects

There are several faint deep-sky objects that can only be seen using larger home telescopes, or from an observatory. They include NGC 625, an irregular dwarf galaxy 12.7 million light-years away, and Robert's Quartet, a compact group of four galaxies that is 160 million light-years away.

Brightest star

Alpha Phoenicis, or Ankaa, is a spectroscopic binary star system. At least one of the stars appears to be a giant, and the mass of the system is estimated to be 2.8 times that of the Sun.

Area and main stars

469 square degrees with 4 main stars

PICTOR

The Painter's Easel

Lore

Lacaille described this constellation in 1756. He originally called it le Chevalet et la Palette (French for 'the easel and palette'). In Chinese astronomy, Pictor lies across one of the quadrants by the Vermilion Bird of the South.

How to see it

Pictor is a southern constellation, visible at latitudes between +26° and -90° and best seen in January. Although small, it is easy to find using the bright star Canopus in Carina (p. 126). Pictor sits between Canopus and the Large Magellanic Cloud.

Deep-sky objects

NGC 1705 is an irregular dwarf galaxy that is 17 million light-years from Earth. This particular galaxy is one of the most actively forming in our nearby universe. It is estimated that, in the past 10 million years, NGC 1705 has formed stars that have a combined mass more than 570,000 times that of the Sun.

Brightest star

Alpha Pictoris is a binary star system with an estimated age of 660 million years. The first star is 1.6 times the mass of the Sun, and the second is approximately the same mass as the Sun.

Area and main stars

247 square degrees with 3 main stars

PUPPIS

The Poop Deck

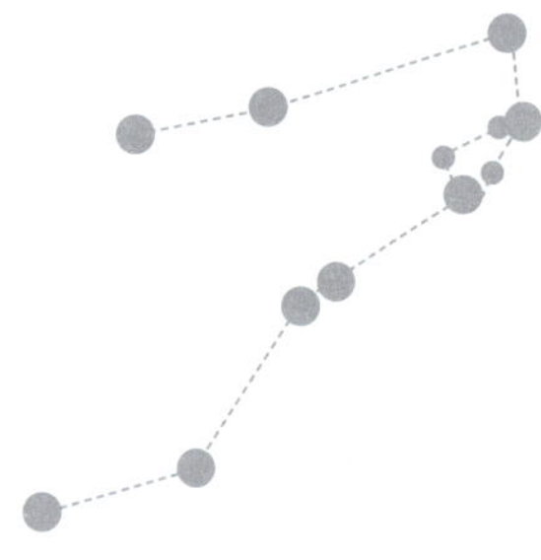

Lore

Puppis was once a part of a larger constellation called Argo Navis, which, in Greek mythology, represented the ship the Argonauts used during their search for the Golden Fleece. The stars that made up Argo Navis were separated into three constellations – Carina, Puppis and Vela – each representing a different part of the ship. Puppis is the poop deck (the elevated area of a ship's deck).

How to see it

✦ Puppis is a southern constellation, visible at latitudes between +40° and -90° and best viewed in February. First, find Vela (p. 192). Puppis sits to the right of Vela's oval shape. It is a rectangular shape made up of nine stars joined together.

Deep-sky objects

Puppis contains the multiple star clusters M46, M47 and M93. It is also home to the Skull and Crossbones Nebula (NGC 2467) and another spectacular planetary nebula, NGC 2440.

Brightest star

⁂ Zeta Puppis is one of just a few O-type stars that can be seen with the naked eye. It is a blue supergiant and one of the most luminous stars in the entire galaxy. It is more than 10,000 times brighter than the Sun. It sits 1080 light-years from Earth.

Area and main stars

673 square degrees with 9 main stars

The Mariner's Compass

Lore

Pyxis is a small constellation that Lacaille named to honour the mariner's compass. The Ancients Greeks once identified the three main stars of Pyxis as the mast of the mythological ship, Argo Navis.

How to see it

Pyxis is a southern constellation, visible at latitudes between $+50^\circ$ and -90° and best seen in March. First, find Orion's belt (p. 86) then head towards Sirius in Canis Major (p. 36), then continue moving beyond Sirius. Go roughly the same distance as the gap between Orion's Belt and Sirius. Pyxis is a faint constellation in that area.

Deep-sky objects

Pyxis is in the plane of the Milky Way, so there are interesting galactic objects to see. NGC 2818 is a planetary nebula 10,400 light-years from Earth that can be seen with medium to large home telescopes. You may also spot the open cluster NGC 2818A, and NGC 2627.

Brightest star

Alpha Pyxidis is a B-type giant that is more than 10 times the mass of the Sun and is 880 light-years from Earth.

Area and main stars

221 square degrees with 3 main stars

RETICULUM

The Reticle

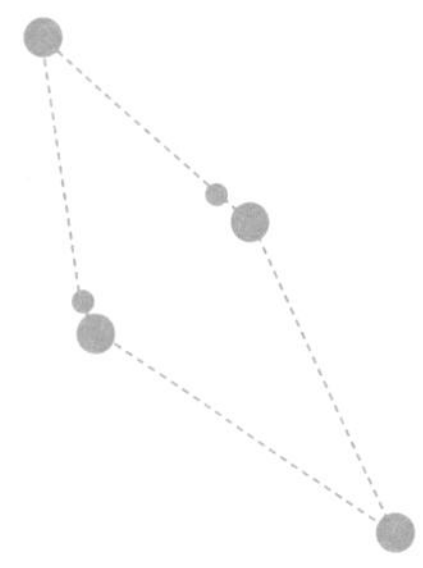

Lore

Reticulum is a small, faint southern constellation that was identified in 1621 by Isaac Habrecht II (1589–1633). He originally named it Rhombus. It was then replaced with a slightly different constellation by Lacaille, who renamed it to commemorate the reticle in a telescope's eye piece. This constellation wasn't officially recognised until 1922.

How to see it

Reticulum is a southern constellation, visible at latitudes between +23° and -90° and best viewed in December. Because it is a small constellation, the stars appear very close together. A night-sky application will be useful. Reticulum sits next to the Large Magellanic Cloud and its four main stars cover an area of the sky similar in size to the cloud.

Deep-sky objects

NGC 1559 is a barred spiral galaxy that can be seen with a medium to large home telescope. Multiple supernovae have been recorded in this galaxy and there appears to be strong star formation occurring.

Brightest star

Alpha Reticuli is 160 light-years from Earth. It is a G-type star that is 3 times the mass of the Sun and only 300 million years old.

Area and main stars

114 square degrees with 4 main stars

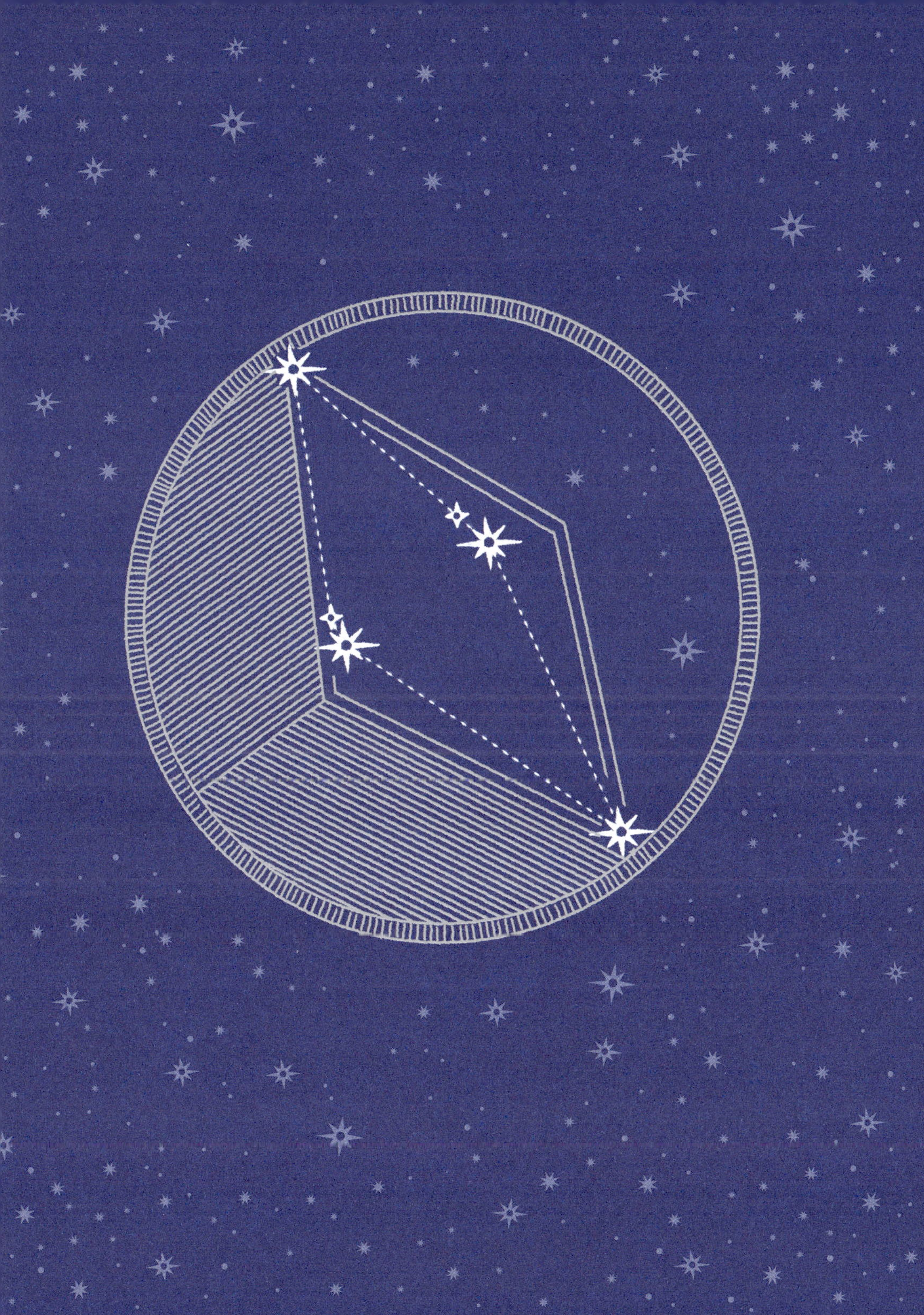

SCULPTOR

The Sculptor

Lore

Sculptor is a faint constellation first described by Lacaille. It was originally named Apparatus Sculptoris ('the sculptors studio'), which was later shortened to just one word. Lacaille depicted it as a three-legged table with a carved head, two chisels and a mallet on it.

How to see it

Sculptor is a southern constellation, visible at latitudes between +50° and -90° and best seen in November. As it is a faint constellation, it is easiest to see away from light-polluted areas. It is bordered by constellations Cetus (p. 48), Fornax (p. 140) and Grus (p. 142). The bright star Fomalhaut in Piscis Austrinus (p. 94) is also nearby. A night-sky application will help you find it.

Deep-sky objects

NGC 253 is also known as the Sculptor Galaxy and the Silver Coin Galaxy. It is 11 million light-years away and is currently undergoing active star formation. Also in this region of the sky is the Cartwheel Galaxy (ESO 350-40), which has been imaged in great detail by the Hubble Space Telescope and the James Webb Space Telescope. It is 500 million light-years from Earth and is known for its unique shape. The Cartwheel Galaxy is best seen through a very large home telescopes or from an observatory.

Brightest star

Alpha Sculptoris is a B-type giant. It is a young star – just 93 million years old – and has a mass 5 times that of the Sun.

Area and main stars

475 square degrees with 4 main stars

SCUTUM

The Shield

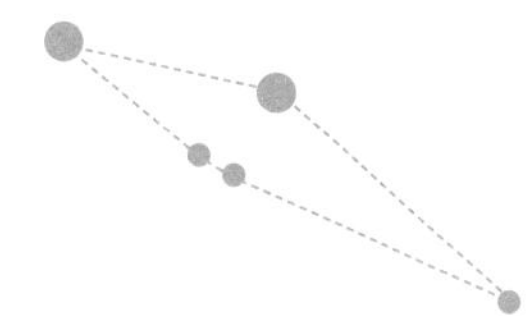

Lore

Scutum was identified in 1687 by Hevelius. He named it Scutum Sobiescianum (Latin for 'Sobieski's shield'), to honour and commemorate the military victories of John III Sobieski, the king of Poland who ruled from 1674 to 1696.

How to see it

Scutum is a southern constellation, visible at latitudes between +80° and -90° and best viewed in August. It is a faint constellation that is easier to see in dark skies. Find Aquila (p. 24) using the bright star Altair on its tail. The head of the bird in Aquila is pointing towards the small diamond shape of Scutum.

Deep-sky objects

Although it is a smaller constellation, Scutum contains many interesting objects, as it sits close to the galactic plane. Viewers can spot multiple open and globular clusters, like M11, M26 and NGC 6712.

Brightest star

Alpha Scuti is a K-type giant. It has a mass 1.33 times that of the Sun and a radius 20 times that of the Sun.

Area and main stars

109 square degrees with 2 main stars

SEXTANS

The Sextant

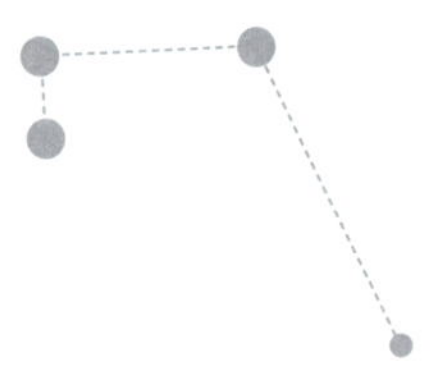

Lore

One of the newest constellations, Sextans was one of the most useful tools astronomers and navigators of the past had. It was first identified by Hevelius and is said to commemorate the sextant that he had lost in a house fire nearly a decade earlier.

How to see it

Sextans is a southern constellation, visible at latitudes between +80° and -90° and best viewed in April. It is an extremely faint constellation, but it can be found with the help of a night-sky application. Find the bright blue star Regulus in Leo (p. 74). Next, look towards the Milky Way to locate the two stars, Alpha Sextantis and Beta Sextantis, that form the middle of the sextant shape.

Deep-sky objects

Sextans is home to the lenticular galaxy NGC 3115, which is 31.6 million light-years from the Milky Way and is the closest galaxy with a known supermassive black hole. Also in this region are the galaxies NGC 3169 and NGC 3166 – spiral galaxies that are on a direct collision course with each other.

Brightest star

Alpha Sextantis is the brightest star of Sextans, but it is faint and is generally only visible on a dark night, far away from the light pollution of cities. It is 2.5 times the mass of the Sun and 385 million years old. Because it is bigger than the Sun, its lifespan will be considerably shorter.

Area and main stars

314 square degrees with 3 main stars

TELESCOPIUM

The Telescope

Lore

Telescopium is a small southern constellation that was first described by Lacaille, who depicted it as an aerial telescope. The telescope was perhaps the most important of all instruments to Lacaille, as it allowed him to catalogue more than 10,000 southern stars during his stay at the Cape of Good Hope. The name was Latinised to Telescopium in 1763.

How to see it

Telescopium is a southern constellation, visible at latitudes between +40° and -90° and best viewed in August. It is a small and faint constellation, best seen away from light-polluted areas. It lies beneath the tail of Scorpius (p. 100).

Deep-sky objects

Within Telescopium is a group of twelve galaxies that are 120 million light-years from the Milky Way. The two brightest galaxies in this group are NGC 6868 and NGC 6861. There is also a globular cluster, NGC 6584, that is 45,000 light-years from Earth.

Brightest star

Alpha Telescopii is a B-type subgiant with a mass 5.2 times that of the Sun. Ptolemy originally included it in the constellation Corona Australis, but it was moved to Telescopium in the 18th century. It is 278 light-years from Earth.

Area and main stars

252 square degrees with 2 main stars

TRIANGULUM AUSTRALE

The Southern Triangle

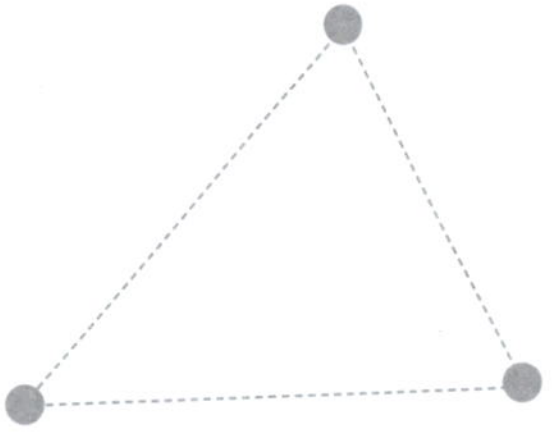

Lore

The first depiction of this constellation was recorded in 1569 by Plancius. The Wardaman people of Australia identified these stars as the tail of the Rainbow Serpent, which is said to stretch between Crux to Scorpius. The Rainbow Serpent is a deity or creator god that features in the stories of many Aboriginal and Torres Strait Island Peoples.

How to see it

Triangulum Australe is a southern constellation, visible at latitudes between +25° and -90° and best viewed in July. Although it is a small constellation, Triangulum Australe is easy to spot thanks to the two bright stars of its neighbouring constellation of Centaurus (p. 44), which can be used as the pointers to Crux (p. 136). When followed in the other direction, they lead to Triangulum Australe.

Deep-sky objects

There are only a few deep-sky objects in this region of the sky. NGC 6025 is an open cluster located 2500 light-years away. NGC 5979 is a planetary nebula that has two stunning rings around it when viewed through large telescopes.

Brightest star

Alpha Trianguli Australis, or Atria, is a K-type supergiant with a mass 7 times that of the Sun. It is 391 light-years from Earth and estimated to be 48 million years old.

Area and main stars

110 square degrees with 3 main stars

TUCANA

The Toucan

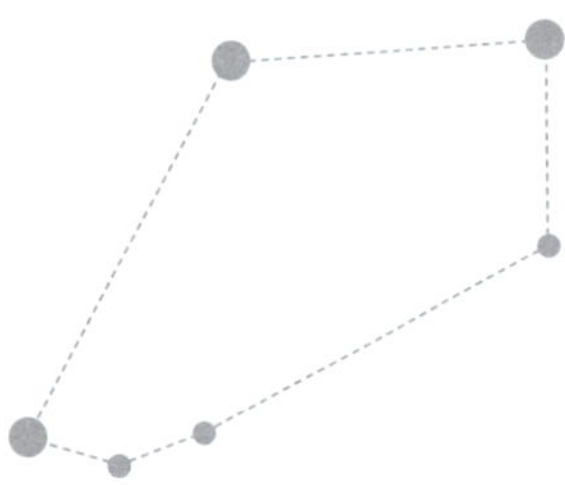

Lore

Tucana was first identified and depicted on a celestial map in 1598 by Plancius. It was named after the toucan, one of the most recognisable birds in the world, due to its long, brightly coloured bill. For a short time, as recorded in a 1603 constellation catalogue, it was named after the Indian magpie.

How to see it

Tucana is a southern constellation, visible at latitudes between +25° and -90° and best viewed in October. First, locate the Large Magellanic Cloud and the Small Magellanic Cloud. Draw a line between them, then extend the line past the Small Magellanic Cloud until its length has doubled. The two stars on either side of the expanded line form the body of Tucana, and its bill branches out from the top star.

Deep-sky objects

Aside from the Small Magellanic Cloud, which is one of the few nearby galaxies visible to the naked eye, viewers can also see the globular cluster NGC 104 and the open cluster NGC 265.

Brightest star

Alpha Tucanae is a binary star system with an orbital period of 11.5 years. The main component of this system is a K-type star. Even with large telescopes, its two stars cannot be resolved individually. They can only be separated by looking at their spectra.

Area and main stars

295 square degrees with 3 main stars

VELA

The Sails

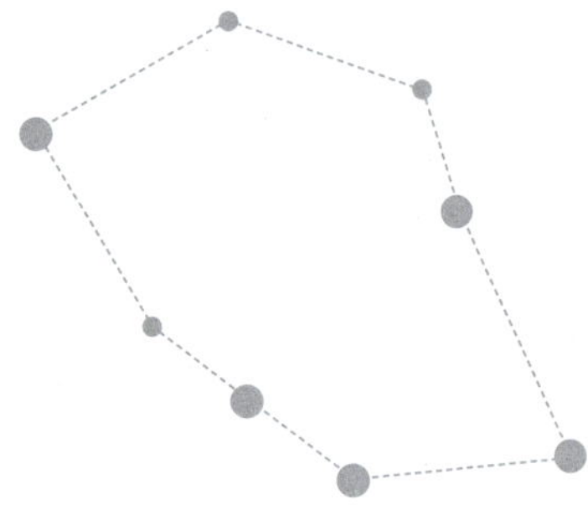

Lore

Vela was once a part of a larger constellation called Argo Navis, which, in Greek mythology, represented the ship the Argonauts used during their search for the Golden Fleece. The stars that made up Argo Navis were separated into three constellations – Carina, Puppis and Vela – each representing a different part of the ship. Vela is the sails.

How to see it

✦ Vela is a southern constellation, visible at latitudes between +30° and -90° and best viewed in March. It appears as an oval shape in the sky. First, locate Crux (p. 136). Next, travel outwards from the lower-middle of Crux for a distance equal to five lengths of Crux. You will arrive at Mu Velorum, a bright star in Vela, which forms the side of the oval.

Deep-sky objects

In this patch of sky, viewers can spot the much-loved Southern Ring Nebula (NGC 3132). This nebula is the subject of the first images from the James Webb Space Telescope, which was launched by NASA in December 2021. The globular cluster NGC 3201 is also in this region.

Brightest star

Gamma Velorum is a quadruple star system that appears as a single star. In telescope images, it appears as just two stars, but both of these stars have another star orbiting them. It's estimated all four stars are less than 5 million years old.

Area and main stars

500 square degrees with 5 main stars

VOLANS

The Flying Fish

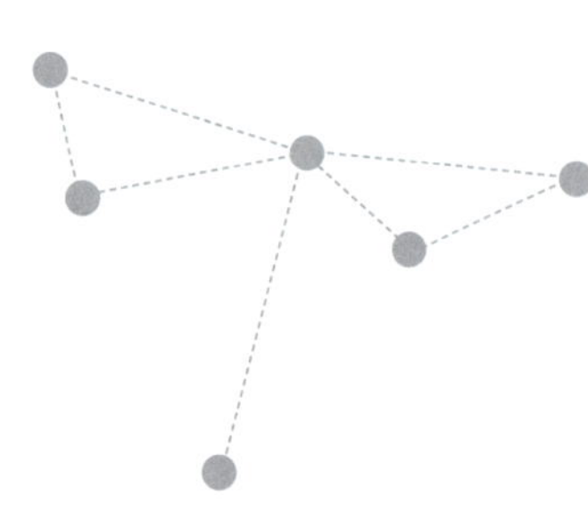

Lore

Volans is one of the twelve constellations introduced by Plancius in 1598. Volans represents a kind of tropical fish that can jump out of the water and appears to fly. In some of the older celestial globes, it is portrayed as a fish being chased by Dorado.

How to see it

Volans is a southern constellation, visible at latitudes between +15° and -90° and best viewed in February. Volans sits beneath the Large Magellanic Cloud and between the constellation of Carina (p. 126) and the south celestial pole.

Deep-sky objects

The Lindsay-Shapley Ring is a stunning ring galaxy located 300 million light-years away. It is a strangely shaped galaxy and, like the Cartwheel Galaxy, is a result of a galaxy collision millions of years ago. Also visible is NGC 2442, a spiral galaxy located 50 million light-years from Earth.

Brightest star

Gamma Volantis is a binary star system composed of a K-type giant and an F-type main sequence star. It is 133 light-years from Earth.

Area and main stars

141 square degrees with 6 main stars

VULPECULA

The Little Fox

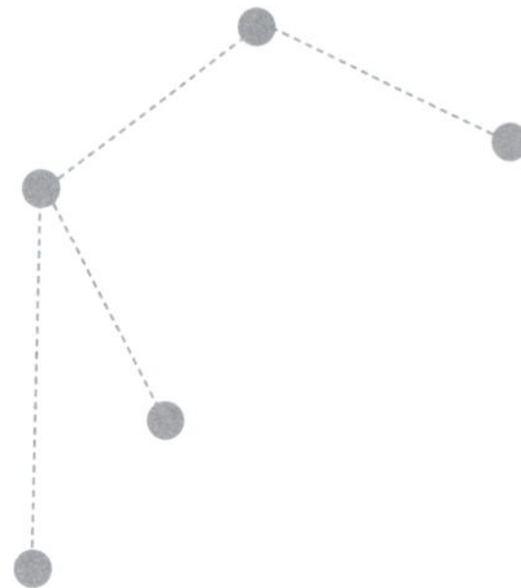

Lore

Vulpecula was introduced by Hevelius in the late 17th century. It was first known as the Little Fox with the Goose, and the original illustrations show a fox holding a goose in its mouth. For a short time, predator and prey were separated into two constellations, but they are now merged into Vulpecula.

How to see it

Vulpecula is a northern constellation, visible at latitudes between +90° and -55° and best seen in September. It is a faint constellation that is best spotted in dark skies away from light pollution. It lies beneath Cygnus (p. 58), near the bright star Vega in Lyra (p. 82).

Deep-sky objects

The large and bright nebula M27, known as the Dumbbell Nebula, can be seen with large binoculars or a small telescopes. It is only 9800 years old and, in 1764, was the first planetary nebula to be discovered. NGC 7052 is an elliptical galaxy located 214 million years from Earth that has a supermassive black hole at its centre.

Brightest star

Alpha Vulpeculae is 291 light-years from Earth. It is a red giant approximately the mass of the Sun but it has expanded to have a radius 43 times that of the Sun, and is estimated to be at least twice as old. It gives us a perfect glimpse into the future of our solar system, when the Sun reaches the end of its life.

Area and main stars

268 square degrees with 5 main stars

Glossary

A-TYPE STAR	hydrogen-burning dwarf star
B-TYPE STAR	hydrogen-burning giant star
BINARY STAR SYSTEM	two stars that are gravitationally bound together and orbit a common centre of mass
BLUE SUPERGIANT	newly evolved and very luminous massive star
DOUBLE-STAR SYSTEM	two stars that lie in same direction from Earth and appear to be very close together, but are not linked by gravity
DARK NEBULA	dense interstellar cloud of gas and dust that can obscure the visible light from objects behind it
ECLIPTIC PLANE	flat, disk-shaped space that lies between the path of an orbiting object (e.g. Earth) and the centre of its orbit (e.g. the Sun)
EXOPLANET	planet that orbits a star other than the Sun
F-TYPE STAR	hydrogen-burning dwarf star
FACE-ON SPIRAL GALAXY	spiral galaxy that appears to face Earth, allowing us to see its spiral structure
G-TYPE STAR	hydrogen-burning dwarf star
GALACTIC ANTICENTER	point directly opposite the galactic centre as viewed from Earth; located in Auriga
GALACTIC CENTRE	central region of the Milky Way galaxy; location of a supermassive black hole and a dense concentration of stars and gas
GALACTIC PLANE	plane across the sky where the majority of the Milky Way's mass lies
GAS GIANT	large planet mainly composed of hydrogen and helium
GLOBULAR CLUSTER	large, compact spherical star cluster; typically contains the oldest stars in a galaxy
GRAND DESIGN SPIRAL GALAXY	large and symmetrical well-organised spiral galaxy
HOT GAS DISK	accretion disk of dust and gas orbiting an astronomical object
LARGE MAGELLANIC CLOUD	dwarf galaxy; satellite neighbour of the Milky Way
LIGHT-YEAR	astronomical distance equivalent to how far light can travel in one year; approximately 9 trillion kilometres
LOCAL GROUP	a group of around fifty galaxies, including the Milky Way, that are bound together by gravity
LOW-MASS STARS	stars with 2 times or less the mass of the Sun
M-TYPE STAR	hydrogen-burning dwarf star
MAIN SEQUENCE STAR	star actively burning hydrogen in its core
MESSIER OBJECT	one of 110 astronomical objects catalogued by French astronomer Charles Messier in the 18th century

MOLECULAR CLOUD	dense, cold and opaque region of space composed of gas and dust
NORTH CELESTIAL POLE	point on the north celestial sphere that the sky appears to rotate around
NORTHERN SKY	region of the sky located in the northern celestial hemisphere
O-TYPE STAR	hydrogen-burning giant star
PLANETARY NEBULA	ring-shaped nebula formed by an expanding shell of gas around an aging star
PRECESSIONAL PATH	slow continuous change in the orientation of the Earth's rotational axis
PROTOPLANETARY DISK	rotating structure of gas and dust surrounding a young, newly formed star
QUADRUPLE STAR SYSTEM	four stars that are gravitationally bound together and orbit a common centre of mass; often made up of two binary star systems
RED CLUMP STAR	star that is slightly hotter than most other stars of the same luminosity
RED GIANT	large star formed when hydrogen is exhausted in the core of a star, and its outer layers expand and cool down
SEYFERT GALAXY	small, compact spiral galaxy with an active black hole at the centre
SMALL MAGELLANIC CLOUD	irregular dwarf galaxy; satellite neighbour of the Milky Way
SOUTH CELESTIAL POLE	point on the south celestial sphere that the sky appears to rotate around
SOUTHERN SKY	region of the sky located in the southern celestial hemisphere
SPECTROSCOPIC BINARY	two stars that are too close together to resolve via imaging, but can be identified as different stars via spectra due to their composition
SPIRAL GALAXY	galaxy with concentrations of stars and gas clouds in rotating spiral arms that curve out from a dense central region
STELLAR MEMBER	star that is gravitationally bound to a group or cluster of stars
SUBGIANT	star that has exhausted the hydrogen fuel in its core and is becoming a giant star
SUPERGIANT	massive blue or red luminous star that is in the later stages of its evolutionary path
TRIPLE STAR SYSTEM	three stars that are gravitationally bound together and orbit a common centre of mass; often made up of a binary star system with a third star orbiting it
VARIABLE STAR	star that appears to change in brightness
WHITE DWARF	dense, hot remnants of a main sequence star that has exhausted its nuclear fuel

Published in 2025 by
Smith Street Books
Naarm (Melbourne) |
Australia
smithstreetbooks.com

Distributed outside of ANZ,
North & Latin America by
Thames & Hudson Ltd.,
6–24 Britannia Street,
London, WC1X 9JD
thamesandhudson.com

EU Authorised
Representative:
Interart S.A.R.L.
19 rue Charles Auray,
93500 Pantin, Paris, France
productsafety@thameshudson.co.uk;
www.interart.fr

ISBN: 978-1-9232-3940-1

Smith Street Books respectfully acknowledges the Wurundjeri People of the Kulin Nation, who are the Traditional Owners of the land on which we work, and we pay our respects to their Elders past and present.

Publisher: Aisling Coughlan
Text: Dr Sara Webb and Dr Kirsten Banks
Illustrations: Aidan Meighan
Design concept: Michelle Macintosh
Design layout and prepress: Megan Ellis
Production manager: Aisling Coughlan

Printed & bound in China by C&C Offset Printing Co., Ltd.

Book 409
10 9 8 7 6 5 4 3 2 1

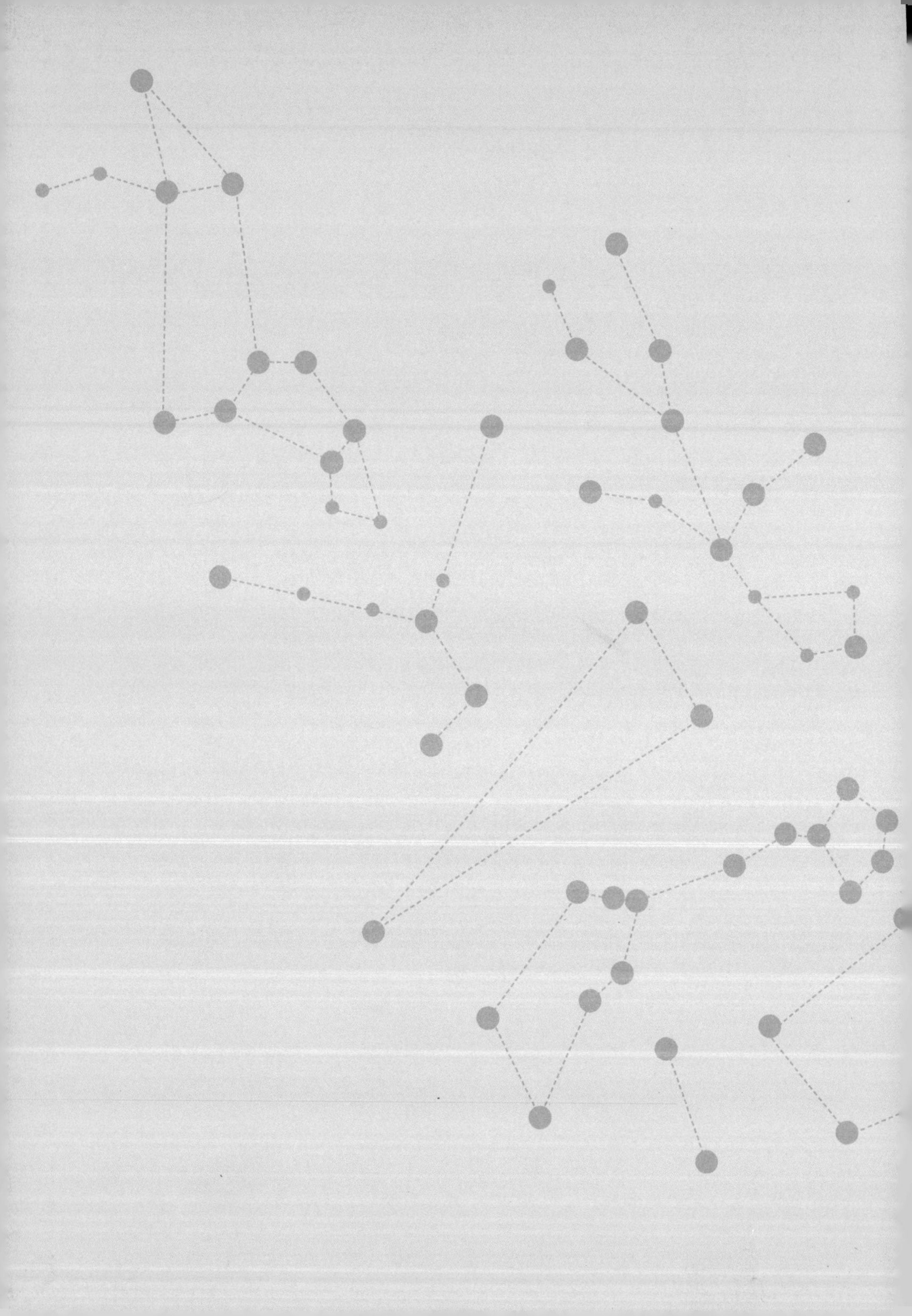